# Modern Geometry

## Second Edition

STEVEN ROMAN

Emeritus Professor
Department of Mathematics
California State University, Fullerton

The Koch Snowflake A Fractal Fern



Ordering: 949-854-5667
7/9/2010



*Modern Geometry*
ISBN 1-878015-25-7

# Preface

This is one in a series of mathematics modules designed for the general college level student, whose background may include only intermediate level algebra. It can be used, together with other modules, or as a supplement to another text, in courses such as liberal arts mathematics, finite mathematics, or discrete mathematics.

Chapter 1 is devoted to the study of polygons, beginning with a section devoted mostly to terminology and the construction of regular polygons. There is a brief discussion of Fermat numbers in this regard, as well as a discussion of star polygons. The next section is devoted to tiling the plane with polygons. Here we discuss and determine all regular and semiregular tilings. The third section concerns polyhedra.

Chapter 2 contains a brief introduction to fractals. We discuss fractals that arise from an iterative process, such as the Koch curve, Sierpinski gasket and Heighway dragon. We also define the similarity dimension of a fractal, which is the only fractal dimension that can be reasonably discussed and computed at this level.

Answers to the odd numbered exercises are given in the back of the module, and answers to all of the exercises are available to the instructor upon request.

# Preface to the Series

This series is designed to provide textbook material for a course in contemporary mathematics for college level students. For one reason or another, large publishers have not responded to the educational concerns of instructors. Through the use of desktop publishing techniques, this series of modules can provide the needed flexibility to adapt to the differing concerns of instructors and motivational needs of students. In particular, by using these modules, instructors can now select topics on a class-by-class basis. We hope that this series will provide both students and instructors with an adaptable learning tool to help increase enthusiasm for mathematics in the classroom.

# Acknowledgments

I would like to express my indebtedness to Professor John G. Pierce for making a number of constructive suggestions for improvements in this module. Also, I am indebted to Ms. Donna Dolan and Ms. Joan Sholars for carefully proofreading the module and working all of the exercises.

# Changes for this Edition

The changes for this edition of this module are strictly cosmetic. We are now preparing our modules in EXP for Windows, Version 5.0. As a result, the page location of items may have changed, but there is no content change.

# Contents

# Chapter 1 Polygons

## 1.1 Polygons and Symmetry

Polygons play a major role in modern geometry. Let us begin our study of these objects with the definition.

**Definition**

Let $P_1, P_2, \ldots, P_n$ be $n$ distinct points in the plane. The collection of line segments $P_1P_2, P_2P_3, \ldots, P_{n-1}P_n$ is called a **polygon**. Each point $P_k$ is called a **vertex** of the polygon and each line segment $P_kP_{k+1}$ is called a **side** of the polygon.★

***Figure 1***

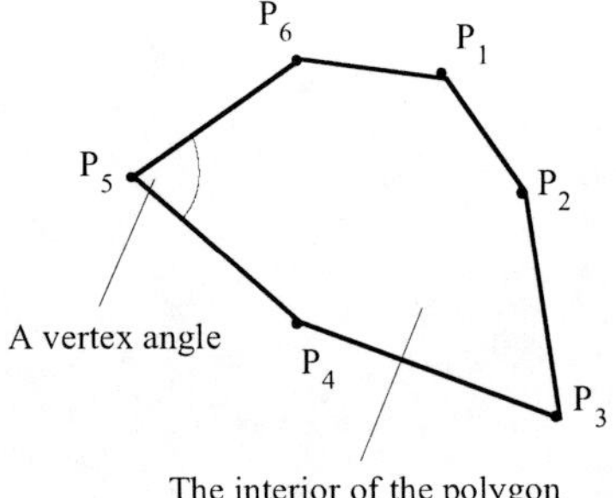


Figure 1 shows an example of a polygon. For emphasis in this figure, we have drawn the vertices as dots. The **interior** of a polygon is the portion of the plane bounded by the polygon, as shown in Figure 1. The **vertex angles** of a polygon are the interior angles formed by adjacent sides of the polygon, as in Figure 1.

Polygons can be classified in several different ways.

**Definition**

1) A polygon is said to be **simple** if its sides do not meet at any points other than the vertices. Otherwise, it is **nonsimple**. Here is an example.

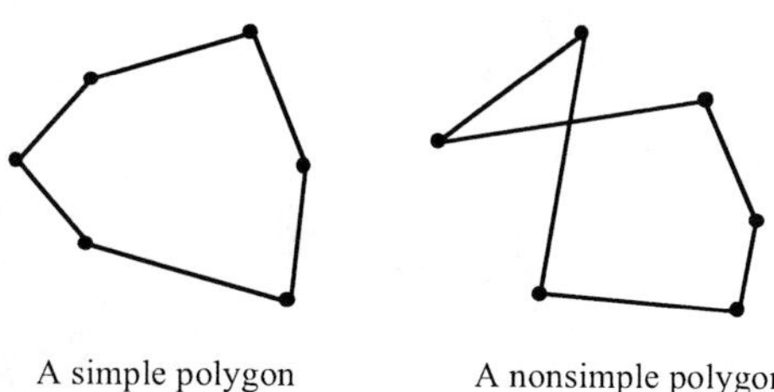

A simple polygon A nonsimple polygon

2) A *simple* polygon is said to be **convex** if whenever the polygon contains two distinct points in its interior, it contains the line segment connecting those two points. Here is an example.

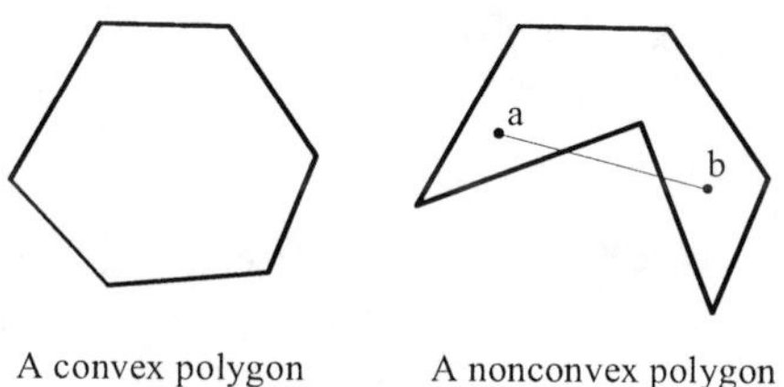


A convex polygon A nonconvex polygon

The polygon on the right is not convex since the line segment connecting points $a$ and $b$ is not contained entirely within the interior of the polygon.★

## Regular Polygons

Regular polygons are probably the most familiar types of polygons.

**Definition**

A *simple* polygon is said to be **regular** if its vertices lie on a circle and all sides have the same length. The center of the circle is called the **center** of the regular polygon. Here is an example.

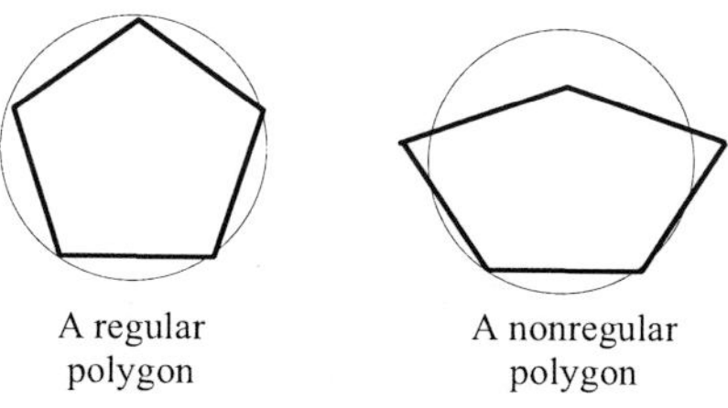

A regular polygon A nonregular polygon

★

The figure below shows some regular polygons. The names in this figure apply to nonregular polygons as well. A regular triangle is called an **equilateral triangle** and a regular quadrilateral is called a **square**.

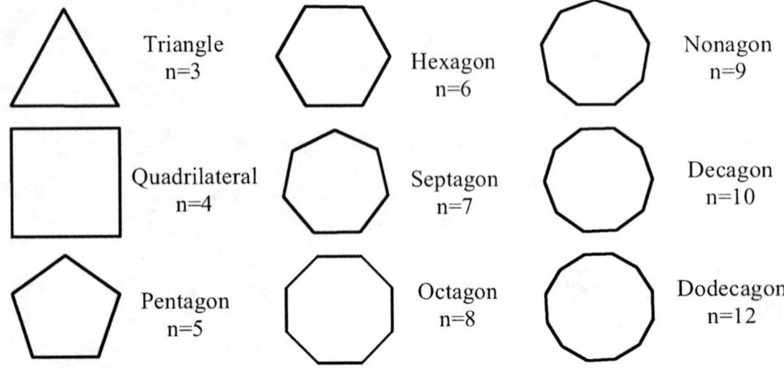


For a *regular* polygon, we can define the **central angles** as shown in Figure 2. It is not hard to see that all of the central angles of a regular polygon are equal.

***Figure 2***

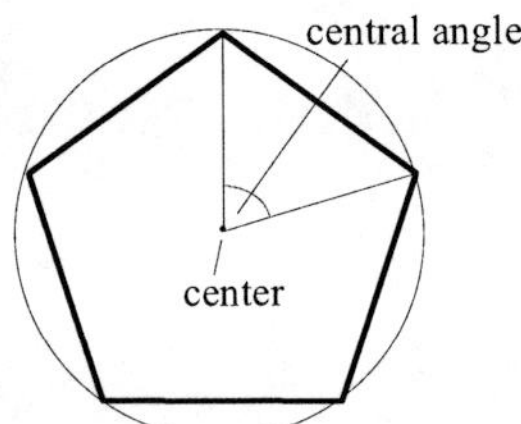


There are simple formulas for the central and vertex angles of a regular polygon, which we describe in the next theorem. We ask you to supply a proof of this theorem in the exercises.

**Theorem 1**

If $P$ is a regular polygon with $n$ sides, then each central angle measures

$$c = \frac{360}{n} \text{ degrees}$$

and each vertex angle measures

$$v = 180\left(1 - \frac{2}{n}\right) \text{ degrees}$$

★

## Construction of Regular Polygons

The question of which regular polygons can be constructed with a straight edge and compass alone is quite fascinating. As early as about 300 B.C. Euclid, in Book IV of the *Elements*, showed how to construct regular polygons with 3, 4, 5, 6 and 15 sides, using only a straight edge and compass. Since it is possible using only these tools to bisect any line segment or angle, once a given polygon is constructed, we can easily construct a regular polygon with twice as many sides. (How?)

Hence, using Euclid's constructions, we can construct regular polygons with any of the following number of sides

$$\begin{array}{l}3, 3\cdot 2, 3\cdot 2^2, 3\cdot 2^3, \ldots\\ 4, 4\cdot 2, 4\cdot 2^2, 4\cdot 2^3, \ldots\\ 5, 5\cdot 2, 5\cdot 2^2, 5\cdot 2^3, \ldots\\ 6, 6\cdot 2, 6\cdot 2^2, 6\cdot 2^3, \ldots\\ 15, 15\cdot 2, 15\cdot 2^2, 15\cdot 2^3, \ldots\end{array}$$

However, it was not until almost the nineteenth century that it was discovered *exactly* which regular polygons can be constructed with straight edge and compass – at least in theory. To understand the situation, we recall that an integer $p > 1$ is said to be **prime** if it has no factors other than 1 and itself. The first few prime numbers are $2, 3, 5, 7, 11, 13, 17, 19, \ldots$ . It is well-known that any integer $n \geq 2$ can be written as a product of prime numbers, in a unique way except for the order of the factors. It is known that there are an infinite number of primes, but no one knows a formula for generating prime numbers. In fact, many of the properties of prime numbers is still unknown.

## Fermat Numbers

A special type of number that plays a role in the construction of regular polygons is the **Fermat number**. A Fermat number is any number of the form

$$f_n = 2^{(2^n)} + 1$$

For instance, the first few Fermat numbers are

$$\begin{aligned}f_0 &= 2^{(2^0)} + 1 = 2^1 + 1 = 3\\ f_1 &= 2^{(2^1)} + 1 = 2^2 + 1 = 5\\ f_2 &= 2^{(2^2)} + 1 = 2^4 + 1 = 17\\ f_3 &= 2^{(2^3)} + 1 = 2^8 + 1 = 257\\ f_4 &= 2^{(2^4)} + 1 = 2^{16} + 1 = 65537\end{aligned}$$

Now, it happens that each of these Fermat numbers, $f_0$, $f_1$, $f_2$, $f_3$ and $f_4$ is prime and Pierre de Fermat (a famous French lawyer and amateur mathematician who lived from 1601 to 1655) himself conjectured that all Fermat numbers were prime. However, we now know that the next Fermat number

$$f_5 = 2^{(2^5)} + 1 = 2^{32} + 1 = 4294967297$$

is not prime.

A great deal of effort has been devoted to determining which Fermat numbers are prime but, as yet, no additional Fermat primes have been discovered other than $f_0, \ldots, f_4$ and many mathematicians believe that no other Fermat numbers are prime. Currently (as of 1992), the only Fermat numbers that have been completely factored are $f_5$, $f_6$, $f_7$, $f_8$, $f_9$ and $f_{11}$, although at least one prime factor has been found for each of the Fermat numbers $f_n$ where $n = 10, 12, 13, 15-21, 23, 25-27, 30$. Hence, we know for sure that these numbers are not prime. Nothing whatever is known, however, about Fermat numbers $f_n$ for $n > 30$.

Now we can return to the discussion at hand. In 1796, Karl Friedrich Gauss, considered to be one of the greatest mathematicians of all time, at the age of 19, proved the following theorem.

**Theorem 2**

A regular polygon with $n$ sides can be constructed using straight edge and compass alone if and only if when $n$ is factored into a product of prime numbers, it has either of the forms

$$n = 2^k p_1 p_2 \cdots p_m \tag{1}$$

or

$$n = 2^k, k \geq 2$$

where the primes $p_1, p_2, \ldots, p_m$ are *distinct* Fermat primes. ★

Here are some examples of Gauss' theorem in action with some small values of $n$.

| | |
|---|---|
| $n = 7$ | Not constructible since 7 is not of the form (1), that is, 7 is not a Fermat prime. |
| $n = 8$ | Constructible since $8 = 2 \cdot 4$ so it can be constructed by "doubling" a square. |
| $n = 9$ | Not constructible since $9 = 3 \cdot 3$ is the product of Fermat primes, but they are *not* distinct! |
| $n = 10$ | Constructible by "doubling" the pentagon. |
| $n = 11$ | Not constructible since it is not a Fermat prime |
| $n = 17$ | Constructible since 17 is a Fermat prime |
| $n = 35$ | Not constructible since $35 = 5 \cdot 7$ but 7 is not a Fermat prime |
| $n = 102$ | Constructible since $102 = 2 \cdot 3 \cdot 17$ and 3 and 17 are Fermat primes. |

Gauss' theorem gives us the complete story on which regular polygons are constructible with straight edge and compass, but only in theory, since no one knows precisely which Fermat numbers are prime. Moreover, even for those regular polygons that we know are constructible, the actual construction can be very difficult. As an illustration of this, a mathematician named Professor Hermes devoted ten years of his life to the problem of constructing a regular polygon with $f_4 = 65537$ sides! In the exercises, we ask you to construct an equilateral triangle, a square, a regular pentagon (with help) and a regular hexagon.

## Star Polygons

A regular polygon is formed by placing $n$ points at equal distances around a circle and connecting adjacent points. If instead we skip points, connecting, for example, every third point, we get a **star polygon**. We will denote a star polygon with $n$ vertices (and hence $n$ sides) in which we connect every $d$th vertex by $\{\frac{n}{d}\}$. The number $d$ is called the **density** of the star polygon. Figure 3 shows a star polygon $\{\frac{5}{2}\}$.

***Figure 3 — The Star*** $\{\frac{5}{2}\}$

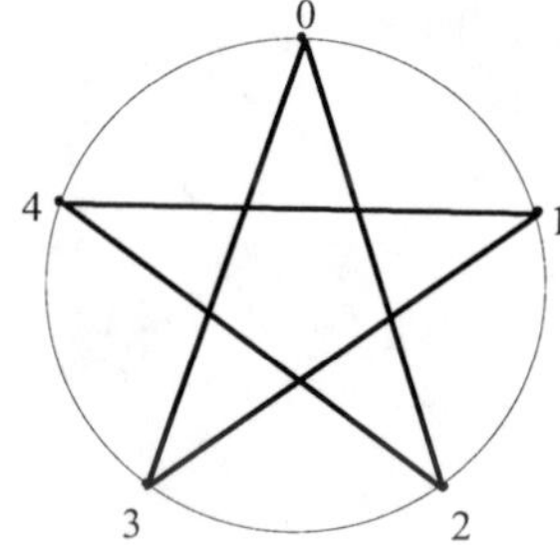


We leave it to you to show that $\{\frac{5}{3}\}$, formed by connecting every 3rd vertex, is the same star polygon as $\{\frac{5}{2}\}$. More generally, $\{\frac{n}{d}\}$ is the same as $\{\frac{n}{n-d}\}$. As a result, we may always assume that $d < n/2$. (What about $d = n/2$?)

Figure 4 shows the star polygons that have been given names, which indicates that these polygons play an important role in geometry.

***Figure 4***

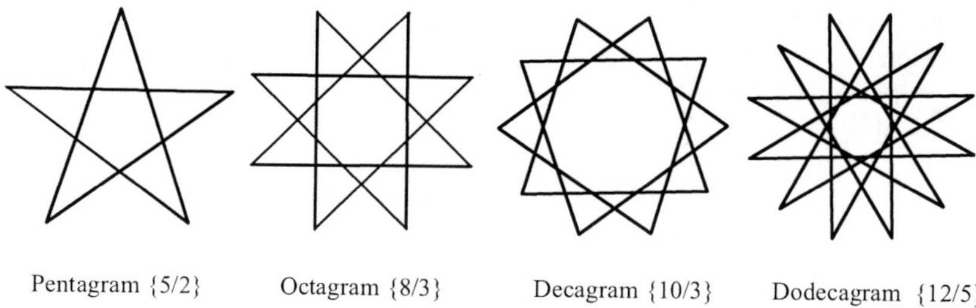

***Figure 5***

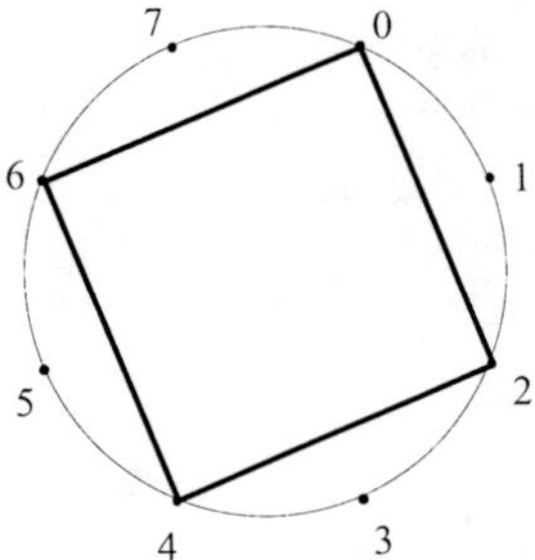


It is not true that a star polygon $\{\frac{n}{d}\}$ exists for every $d < n/2$. For instance, if $n = 8$ there is no star polygon $\{\frac{8}{2}\}$, for if we try to form such a polygon as shown in Figure 5, we come back to the starting vertex (labeled 0) before reaching all vertices. However, as Figure 4 shows, we can construct the octogram $\{\frac{8}{3}\}$. Thus, $\{\frac{8}{3}\}$ exists but $\{\frac{8}{2}\}$ does not exist.

This raises the question of which values of $d$ does there exist a star polygon $\{\frac{n}{d}\}$. The answer is given in the next theorem. First we need a definition. Two positive integers $a$ and $b$ are **relatively prime** if they have no common factors other than 1. For instance, 3 and 8 are relatively prime, but 6 and 8 are not relatively prime, since they have a common factor of 2.

**Theorem 3**

A star polygon $\{\frac{n}{d}\}$ exists if and only if

1) $2 \leq d \leq n - 2$
2) $d$ and $n$ are relatively prime. ★

This theorem explains why $\{\frac{8}{2}\}$ does not exist, for 2 and 8 are not relatively prime (having the common factor 2). It also explains why $\{\frac{8}{3}\}$ exists, for 3 and 8 are relatively prime.

## EXERCISES

1. Verify that 641 is a factor of $f_5$.
2. Draw a picture of a pentagon that has 3 collinear vertices.
3. Verify the formulas in Theorem 1. *Hint*: Referring to the figure below, explain why the triangles formed by drawing radial lines from the center of the polygon to its vertices are congruent. Use this fact to determine the central angle $c$. Find a formula relating $c$ and $d$ and then solve for $d$.

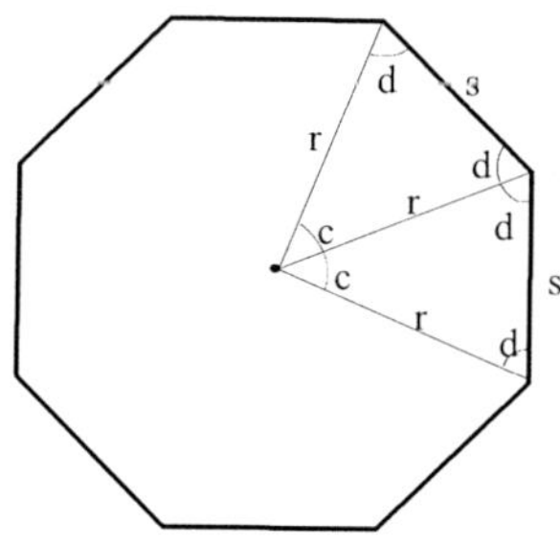


*Let P be a polygon and let L be a line through P. Then L divides P into two pieces. If these pieces are mirror images of one another, we say that L is a **line of symmetry** for P. For instance, a hexagon has six different lines of symmetry, as shown below*

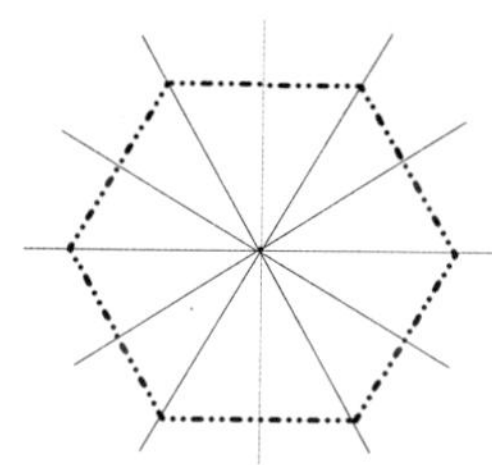

Reflectional Symmetry
of the Hexagon

4. Determine the number of lines of symmetry of an equilateral triangle and draw a picture to support your conclusions.
5. Determine the number of lines of symmetry of a square and draw a picture to support your conclusions.
6. Determine the number of lines of symmetry of a regular pentagon and draw a picture to support your conclusions.
7. How many lines of symmetry does a regular $n$-gon have? Describe these lines.
8. Draw a picture of a polygon that has no lines of symmetry.

*A polygon P has **rotational symmetry through an angle $\theta$ with center of rotation A** if when the polygon is rotated about the point A clockwise through an angle of $\theta$ degrees, the resulting figure exactly fits over the original polygon. For example, a square has rotational symmetry through an angle of 90°, 180°, 270° or 360°, with center of rotation equal to the center of the square.*

9. If a polygon has rotational symmetry through an angle $\theta$, then does it also have rotational symmetry through an angle $\theta + 360°$? Justify your answer. Through what other angles does it have rotational symmetry?
10. Find all of the angles of rotational symmetry of an equilateral triangle. Express them as multiples of a certain angle.
11. Find all of the angles of rotational symmetry of a square. Express them as multiples of a certain angle.

12. Find all of the angles of rotational symmetry of a regular pentagon. Express them as multiples of a certain angle.
13. For a *regular* $n$-gon $P$, find an angle $\alpha$ such that $P$ has rotational symmetry through an angle $\theta$, with center equal to the center of $P$, if and only if $\theta$ is a multiple of $\alpha$. This angle $\alpha$ is called the **fundamental angle of rotation** for $P$. Express $\alpha$ in terms of $n$.
14. Draw a polygon that has no rotational symmetry through an angle less than 360°.
15. Let $P$ be any polygon and let $C$ be a center of rotation. (In this problem, all rotation is with center $C$.)
    a) Show that if $P$ has rotational symmetry through an angle $\theta$ and also through an angle $\phi$ where $\phi < \theta$, then it has rotational symmetry through the angle $\theta - \phi$.
    b) Suppose that $\alpha$ is the smallest angle through which $P$ has rotational symmetry. Use the results of part a) to argue that $P$ has rotational symmetry through an angle $\theta$ if and only if $\theta$ is a multiple of $\alpha$. As in the previous problem, $\alpha$ is called the **fundamental angle of rotation**. *Hint*: suppose $P$ has rotational symmetry through $\theta$. Then by part a), it also has rotational symmetry through the angles $\theta - \alpha$, $\theta - 2\alpha$, $\theta - 3\alpha$,... . What must happen to these angles eventually?
16. Construct a regular (equilateral) triangle using only straight edge and compass.
17. Construct a square using only straight edge and compass.
18. Construct a regular hexagon using only straight edge and compass. *Hint*: start by constructing an equilateral triangle.
19. Construct a regular pentagon using only straight edge and compass. *Hint*: First draw a circle with center $C$. Label any point on the circle $P_1$ and draw a radius through $CP_1$. Then draw a radius through $C$ and a point $A$ that is perpendicular to the first radius. Find the midpoint $M$ of $CA$ and connect it to $P_1$. Next bisect angle $CMP_1$ and label the point where the bisector intersects the radius $CP_1$ with the label $B$. Construct a perpendicular to $CP_1$ through $B$. Label the point where this perpendicular intersects the circle with $P_2$. Then the line segment $P_1P_2$ is one side of the pentagon.
20. Suppose that $r$ and $s$ are integers greater than 2 and that $r$ and $s$ have no common factors other than 1. (That is, $r$ and $s$ are relatively prime.) Show that if a regular $r$-gon and a regular $s$-gon are both constructible with straight edge and compass, then so is a regular $rs$-gon. *Hint*: You may use the fact that, because $r$ and $s$ are relatively prime, there must exist integers $a$ and $b$ (not necessarily positive) for which $ar + bs = 1$. The fact that a regular $r$-gon is constructible means that the angle $\frac{360}{r} = \frac{360}{rs}s$ is constructible. If an angle is constructible, so is any multiple of that angle. If two angles are constructible then so is their sum and their difference. Put all of these pieces together to solve this problem.
21. Suppose that $n = rs$ where $r$ and $s$ are integers greater than 2. Show that if a regular $n$-gon is constructible, then so is a regular $r$-gon and a regular $s$-gon. *Hint*: Draw a regular 12-gon (dodecagon) by hand. (It doesn't have to be perfect.) Note that $12 = 3 \cdot 4$. How would you construct a square and a triangle using the vertices of this dodecahedron?

22. Determine which regular polygons with at most 100 sides are constructible with straight edge and compass.
23. Show that $\{\frac{5}{3}\}$ is the same star polygon as $\{\frac{5}{2}\}$.
24. Show that each vertex angle of a star polygon $\{\frac{n}{d}\}$ has measure

$$180\left(1-\frac{2d}{n}\right)$$

How does this formula compare with the formula for a vertex angle of a regular polygon given in Theorem 1?
25. Use Theorem 3 to find all $d$ for which there exists a star polygon $\{\frac{12}{d}\}$.
26. Use Theorem 3 to find all $d$ for which there exists a star polygon $\{\frac{36}{d}\}$.
27. If $n$ is a prime number, how many star polygons $\{\frac{n}{d}\}$ are there?

# 1.2 Tiling the Plane with Polygons

In this section, we will discuss the question of how to tile the plane with polygons. By a **tiling** (also called a **tessellation**) of the plane, we mean a laying down of *nonoverlapping* geometric shapes in such a way that every point in the plane is covered. It is possible to tile the plane in an infinite variety of ways, and the concept of a tiling was made popular by the Dutch artist M. C. Escher. Figure 1 shows an example of a tiling of the plane using "chickens."

***Figure 1***

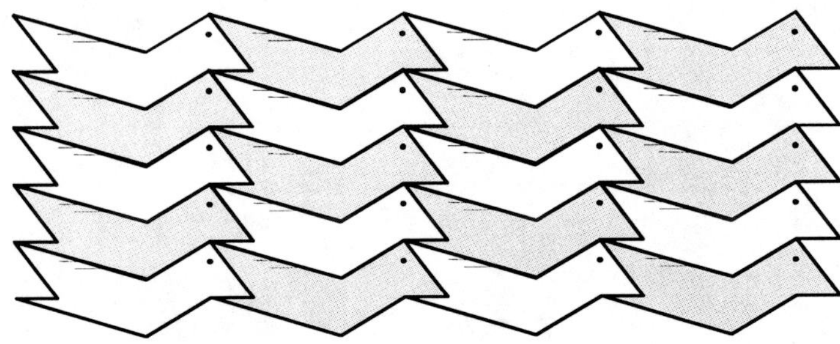

Our interest will center on tiling the plane with polygons, especially regular polygons. First, however, we should set some terminology. Figure 2 shows two tilings of the plane using squares.

***Figure 2***

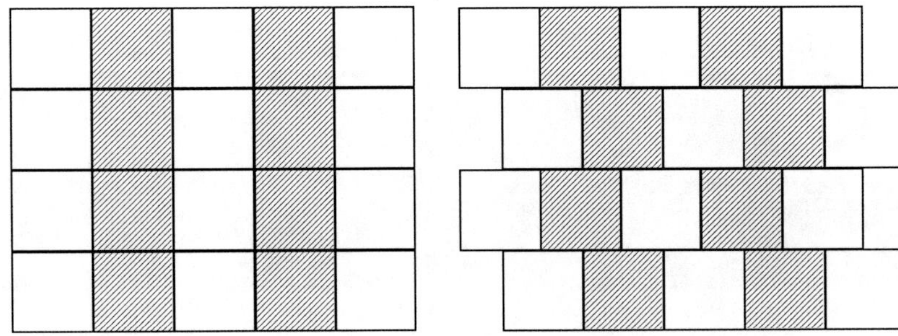

An edge-to-edge tiling

A tiling that is not edge-to-edge

The tiling on the left is said to be **edge-to-edge** because the edges of each tile precisely meet the edges of adjacent tiles. On the other hand, the tiling on the right is not edge-to-edge. From now on, we will restrict our attention to edge-to-edge tilings.

The edges of an edge-to-edge tiling meet at points called the **vertices** of the tiling. The patterns that are formed at these vertices are called the **vertex figures** of the tiling. Figure 3 shows a tiling by regular polygons that has two different types of vertex figures.

*Figure 3*

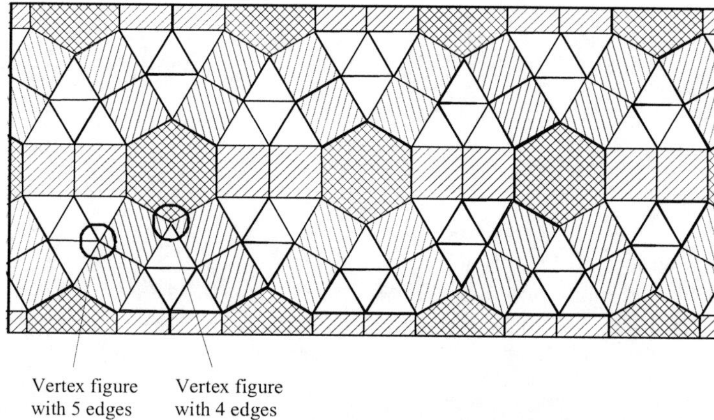


An edge-to-edge tiling of the plane by *regular* polygons that has only one type of vertex figure is referred to as a **semiregular tiling**. If in addition, the tiling uses only one type of regular polygon we say that it is a **regular tiling**. Figure 4 shows a tiling that is semiregular, since each vertex figure is the same, but not regular, since it uses more than one type of tile.

*Figure 4*

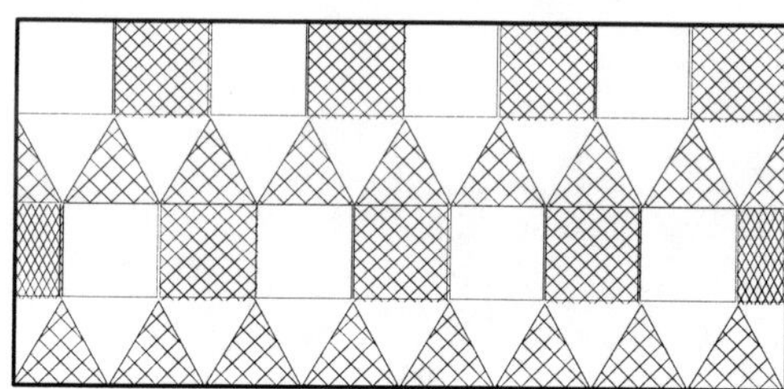

A semiregular tiling that is not regular

Our plan is to determine all possible regular and semiregular tilings.

## Regular Tilings of the Plane

***Figure 5***

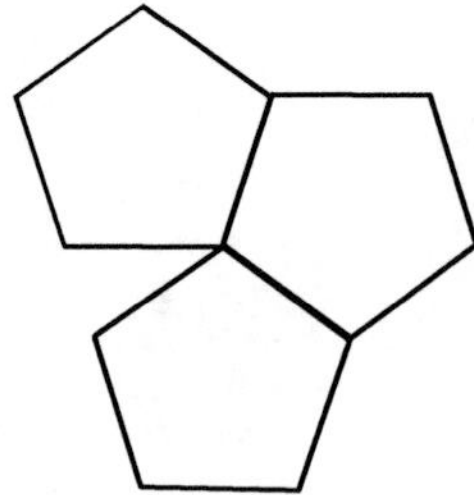

A regular tiling of the plane is one that uses only one regular polygon, of one size. Not all regular polygons can be so used to tile the plane however. For instance, Figure 5 shows what happens when we try to tile using a regular pentagon. Since the vertex angle of a pentagon is

$$v = 180\left(1 - \frac{2}{5}\right) = 180 \cdot \frac{3}{5} = 108°$$

we see that 3 pentagons placed at a vertex subtend an angle of only $3 \cdot 108 = 324° < 360°$, whereas $4$ pentagons subtend an angle of $4 \cdot 108° = 432° > 360°$.

More generally, if a set of identical regular polygons are going to "fit together" at a vertex, then the vertex angle of each polygon must divide $360$. Moreover, the vertex angle cannot exceed 120°, since there must be at least 3 polygons at each vertex. Using the formula for the vertex angle of a regular polygon, we prepare the following table.

| **Table 1** | |
|---|---|
| Number of sides | Vertex angle |
| 3 | 60° |
| 4 | 90° |
| 5 | 108° |
| 6 | 120° |
| 7 | $128\frac{4}{7}°$ |
| 8 | 135° |
| 9 | 140° |
| 10 | 144° |
| 11 | $147\frac{3}{11}°$ |
| 12 | 150° |

An $n$-gon with $n > 6$ will have vertex angle larger than 120°, and so could not be used for a regular tiling of the plane. Also, since 108° does not divide 360°, a pentagon cannot be used for a regular tiling. This leaves us with only 3 possibilities – an equilateral triangle, a square and a regular hexagon. As it turns out, each of these can indeed be used to produce a regular tiling of the plane, as shown in Figure 6.

***Figure 6***

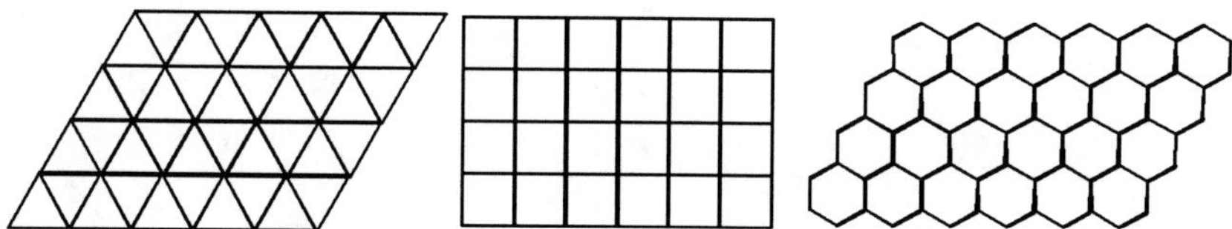

The 3 regular tilings of the plane - by equilateral triangles, squares and hexagons

Thus, we can say that *there are exactly 3 regular tilings of the plane, as shown in Figure 6.*

## Semiregular Tilings of the Plane

It was not difficult to determine all the regular tilings of the plane. To determine all the semiregular tilings, we must do considerably more work. The first step is to recognize that the number of tiles that meet at each vertex must lie between 3 and 6. There cannot be more than 6 tiles since each tile must have vertex angle at least 60° and $7 \cdot 60° = 420° > 360°$.

Let us first think about possible semiregular tilings with 3 tiles at each vertex. If the tiles have $n_1$, $n_2$ and $n_3$ sides then, using the formula for vertex angles, we must have

$$180\Big(1-\frac{2}{n_1}\Big)+180\Big(1-\frac{2}{n_2}\Big)+180\Big(1-\frac{2}{n_3}\Big)=360$$

This simply says that the sum of the 3 vertex angles is 360°. This formula can be simplified as follows

$$\begin{aligned}\Big(1-\frac{2}{n_1}\Big)+\Big(1-\frac{2}{n_2}\Big)+\Big(1-\frac{2}{n_3}\Big)&=2\\ 3-2\Big(\frac{1}{n_1}+\frac{1}{n_2}+\frac{1}{n_3}\Big)&=2\\ \frac{1}{n_1}+\frac{1}{n_2}+\frac{1}{n_3}&=\frac{1}{2}\end{aligned}$$

In a similar manner, if the tiling has 4 tiles at each vertex, we get

$$180\Big(1-\frac{2}{n_1}\Big)+180\Big(1-\frac{2}{n_2}\Big)+180\Big(1-\frac{2}{n_3}\Big)+180\Big(1-\frac{2}{n_4}\Big)=360$$

which simplifies as follows

$$\begin{aligned}\Big(1-\frac{2}{n_1}\Big)+\Big(1-\frac{2}{n_2}\Big)+\Big(1-\frac{2}{n_3}\Big)+\Big(1-\frac{2}{n_4}\Big)&=2\\ 4-2\Big(\frac{1}{n_1}+\frac{1}{n_2}+\frac{1}{n_3}+\frac{1}{n_4}\Big)&=2\\ \frac{1}{n_1}+\frac{1}{n_2}+\frac{1}{n_3}+\frac{1}{n_4}&=1\end{aligned}$$

We leave it as an exercise for you to derive the corresponding formulas for 5 tiles and 6 tiles. Here is the complete list

$$\frac{1}{n_1}+\frac{1}{n_2}+\frac{1}{n_3}=\frac{1}{2} \tag{1}$$

$$\frac{1}{n_1}+\frac{1}{n_2}+\frac{1}{n_3}+\frac{1}{n_4}=1 \tag{2}$$

$$\frac{1}{n_1}+\frac{1}{n_2}+\frac{1}{n_3}+\frac{1}{n_4}+\frac{1}{n_5}=\frac{3}{2} \tag{3}$$

$$\frac{1}{n_1}+\frac{1}{n_2}+\frac{1}{n_3}+\frac{1}{n_4}+\frac{1}{n_5}+\frac{1}{n_6}=2 \tag{4}$$

Our goal is to find all solutions to (1) – (4) in integers $n_i \geq 3$. One way to find all such solutions is to use a computer to do an exhaustive search. However, since we have no upper bound on the sizes of $n_1$, $n_2$ or $n_3$, there are infinitely many possibilities! We must find a way to limit the number of possibilities to a finite number before we can enlist the aid of a computer.

Table 1 will help us do just that. Suppose that we wish to match a "large" tile $T$ with several smaller or equal size tiles. Let the large tile $T$ have vertex angle $t$. Thus, the sum of the vertex angles of the other tiles is $s = 360 - t$. Since $t < 180$, we have $s = 360 - t > 180$.

Now let us make a list of the possible sums of vertex angles for the smaller tiles, up to and including the smallest sum just larger than 180°.

| | |
|---|---|
| 2 terms: | $60 + 60 = 120$ |
| | $60 + 90 = 150$ |
| | $60 + 108 = 168$ |
| | $60 + 120 = 180$ |
| | $60 + 128\frac{4}{7} = 188\frac{4}{7}$ |
| 3 terms: | $60 + 60 + 60 = 180$ |

There are no combinations of 4 or more terms that sum to less than $188\frac{4}{7}$, since $4 \cdot 60 = 240 > 188\frac{4}{7}$.

Our computations show that the sum $s$ of vertex angles *cannot* lie in the range

$$180 < s \leq 188$$

and so must actually be larger than 188. But then

$$t = 360 - s < 360 - 188 = 172$$

Substituting an angle of 172° into the formula for the vertex angle of a regular $n$-gon, we get

$$
\begin{aligned}
180\left(1-\frac{2}{n}\right) &= 172\\
1-\frac{2}{n} &= \frac{172}{180}\\
\frac{2}{n} &= \frac{8}{180}=\frac{2}{45}\\
n &= 45
\end{aligned}
$$

Thus, we have shown that the integers $n_i$ in formulas (1) – (4) must be less than 45, and so must lie in the range $3 \leq n_i \leq 44$.

We can make the computer's work easier by ruling out some additional cases. For instance, the vertex angle of a regular 44-gon is $171\frac{9}{11}°$ and it is not hard to see that such a vertex angle cannot be used in a semiregular tiling. A similar statement holds for 43-gons. There is a possibility using a 42-gon, however, namely

$$\frac{1}{42}+\frac{1}{7}+\frac{1}{3}=\frac{21}{42}==\frac{1}{2}$$

Now, a simple computer program, which takes about 15 minutes to run on a personal computer, yields the following 21 solutions to equations (1)–(4), lying in the range $3 \leq n_i \leq 44$

| **Table 2** | | | | | | | |
|---|---|---|---|---|---|---|---|
| Solution No | $n_1$ | $n_2$ | $n_3$ | $n_4$ | $n_5$ | $n_6$ | |
| 1 | 3 | 3 | 3 | 3 | 3 | 3 | regular |
| 2 | 4 | 4 | 3 | 3 | 3 | | semiregular |
| 2' | 4 | 3 | 4 | 3 | 3 | | semiregular |
| 3 | 6 | 3 | 3 | 3 | 3 | | semiregular |
| 4 | 4 | 4 | 4 | 4 | | | regular |
| 5 | 6 | 4 | 4 | 3 | | | neither |
| 5' | 6 | 4 | 3 | 4 | | | semiregular |
| 6 | 6 | 6 | 3 | 3 | | | neither |
| 6' | 6 | 3 | 6 | 3 | | | semiregular |
| 7 | 12 | 4 | 3 | 3 | | | neither |
| 7' | 12 | 3 | 4 | 3 | | | neither |
| 8 | 6 | 6 | 6 | | | | regular |
| 9 | 8 | 8 | 4 | | | | semiregular |
| 10 | 10 | 5 | 5 | | | | neither |
| 11 | 12 | 6 | 4 | | | | semiregular |
| 12 | 12 | 12 | 3 | | | | semiregular |
| 13 | 15 | 10 | 3 | | | | neither |
| 14 | 18 | 9 | 3 | | | | neither |
| 15 | 20 | 5 | 4 | | | | neither |
| 16 | 24 | 8 | 3 | | | | neither |
| 17 | 42 | 7 | 3 | | | | neither |

Notice that solution $2'$ is the same solution as solution 2 except that the numbers appear in a different arrangement. This results in a different vertex figure and hence a different tiling. Figure 7 shows the two different vertex figures for these solutions. The same applies to solutions $5'$, $6'$ and $7'$.

***Figure 7***

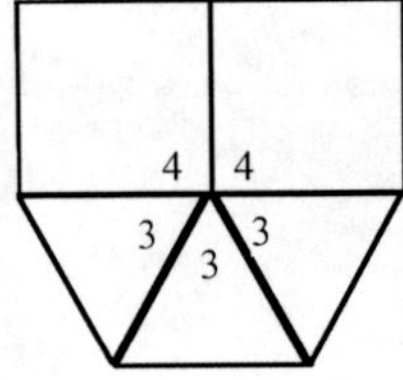


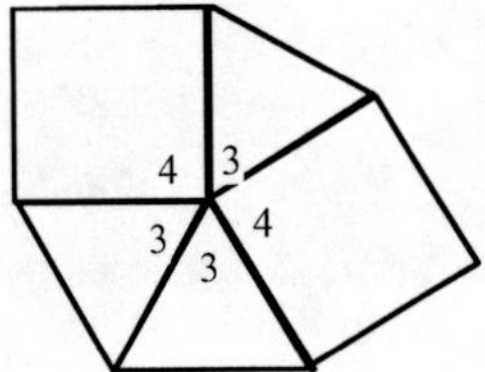


Note also that certain rearrangements of the numbers in a solution are *not* considered to yield different vertex figures. For instance, the solutions 12, 4, 3, 3 and 12, 3, 3, 4 give the same vertex figure. (One is the same as the other when read counterclockwise instead of clockwise.)

Each of the solutions in Table 2 must be checked to see if it yields a semiregular tiling of the plane. After all, just because a set of tiles fits together at one vertex does not mean that it will fit together at all vertices to produce the same vertex figure. For instance, the solutions 7 or $7'$ can be used to produce a tiling of the plane, but not a semiregular tiling. When this is done, we find that *there are exactly 8 semiregular tilings of the plane, as shown below.* The situation for each solution is given in Table 2.

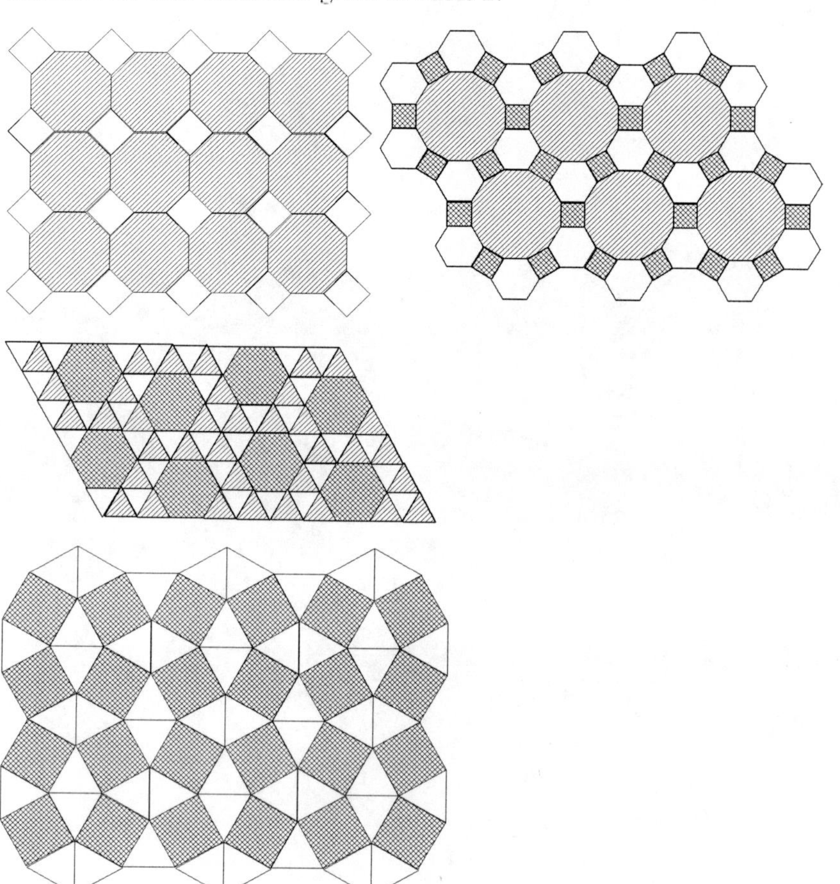

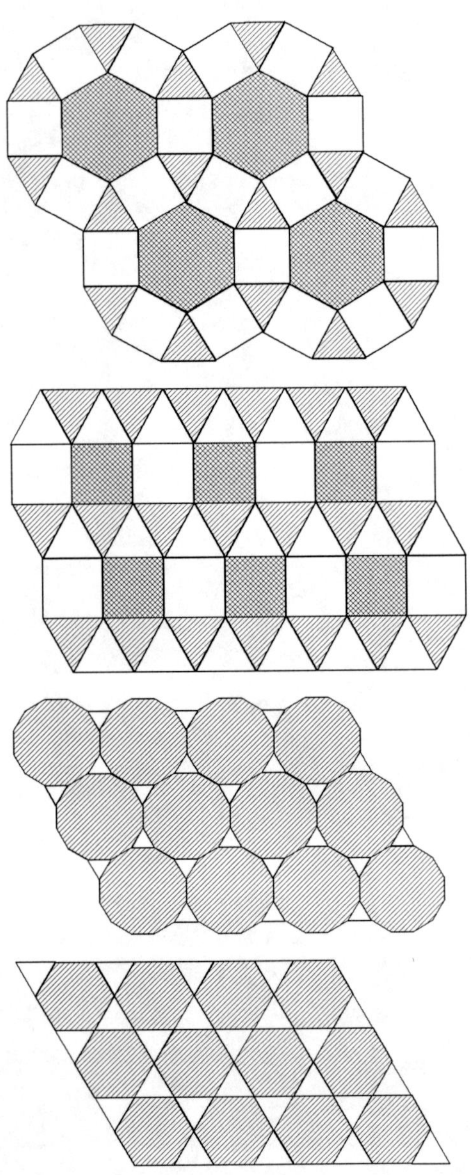

## EXERCISES

1. Define the terms regular tiling, semiregular tiling, and edge-to-edge tiling.
2. Derive formula 3.
3. Derive formula 4.
4. We have seen that a regular (equilateral) triangle will tile the plane. Show that *any* triangle will tile the plane. *Hint*: show that any parallelogram can tile the plane.
5. We have seen that a regular quadrilateral (that is, a square) will tile the plane. Show that *any* quadrilateral will tile the plane. *Hint*: describe how the

partial tiling shown below was constructed using the quadrilateral in the center.

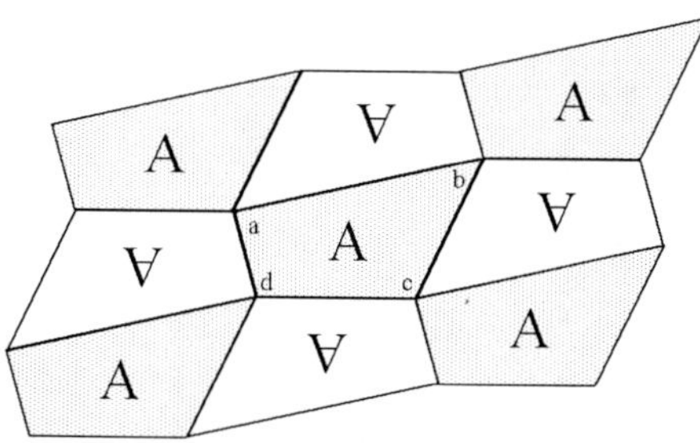


Explain how this can be extended as far as desired, thus tiling the entire plane. Does this work for any quadrilateral?

6. We have seen that a regular pentagon will not tile the plane by itself. However, certain nonregular pentagons can tile the plane. Begin by showing how to tile the plane with the hexagon shown below. Then tile the hexagon into 4 congruent pentagons to obtain a tiling of the plane with pentagons only.

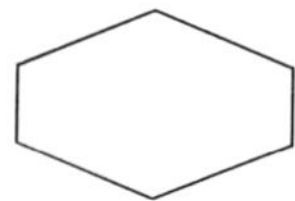

7. Show that a regular 43-gon cannot be used in a semiregular tiling of the plane.
8. Show that a regular 44-gon cannot be used in a semiregular tiling of the plane.
9. Draw the vertex figures corresponding to the solutions 5 and 5′ from Table 2.
10. Draw the vertex figures corresponding to the solutions 6 and 6′ from Table 2.
11. Draw the vertex figures corresponding to the solutions 7 and 7′ from Table 2.
12. An equilateral triangle is **replicating** since several (in this case 3) congruent equilateral triangles can be put together to form a larger equilateral triangle. Find another replicating regular polygon. Are there any others?
13. Draw a vertex figure for solution 7 of Table 2. Then give a convincing argument why you cannot extend this figure to produce a *semiregular* tiling of the plane.
14. Draw a vertex figure for solution 7′ of Table 2. Then give a convincing argument why you cannot extend this figure to produce a *semiregular* tiling of the plane.
15. Consider solutions 13, 14, 16 and 17 of Table 2. In each of these cases, a tiling would involve an equilateral triangle surrounded by 3 other tiles from

among two *different* regular polygons of larger size. Explain why it is impossible to make such an arrangement that will produce the same vertex figure at each vertex of the triangle. Hence, none of these solutions yield semiregular tilings.

16. Give a convincing argument to show why solution 10 of Table 2 does not yield semiregular tilings. *Hint*: consider the arrangement of tiles around one of the pentagons.
17. Give a convincing argument to show why solution 15 of Table 2 does not yield semiregular tilings. *Hint*: consider the arrangement of tiles around the pentagon.
18. Explain why solution 5 of Table 2 cannot give rise to a semiregular tiling of the plane.
19. Explain why solution 6 of Table 2 cannot give rise to a semiregular tiling of the plane.
20. The **dual** of a semiregular tiling is the tiling obtained by connecting the centers of touching tiles. Draw the dual tilings to the 3 regular tilings. What do you get?
21. Referring to the previous exercise, pick a semiregular tiling and draw its dual tiling.

# 1.3 Polyhedra

We now turn our attention to geometric figures in 3-dimensional space. A **polyhedron** (*pl.* polyhedra) is a union of polygonal regions that encloses a bounded region in 3-dimensional space. Figure 1 is an example of a polyhedron. Each of the polygonal regions is called a **face** of the polyhedron. The line segments that make up the faces are called the **edges** and the endpoints of the edges are called the **vertices** of the polyhedron.

***Figure 1 — A polyhedron***

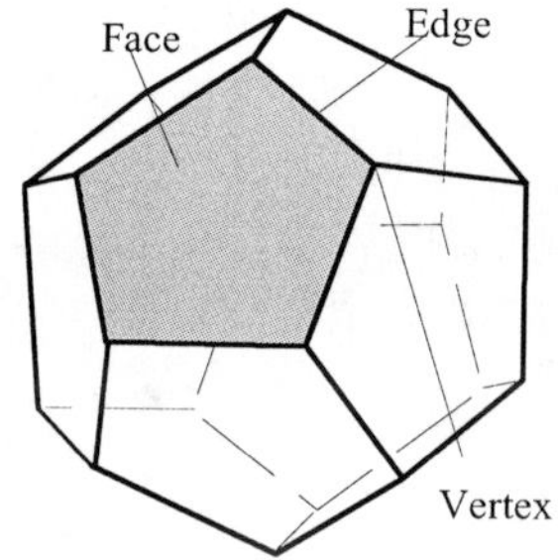


The polyhedron in Figure 1 is **convex** since for any two points inside the polyhedron, the line segment connecting those points also lies inside the polyhedron. We will consider only convex polyhedra in this section.

Prisms and Pyramids

Figure 2 shows some examples of polyhedra known as prisms. A **prism** is a polyhedron formed by taking two congruent polygons, placing them in parallel but distinct planes and then connecting corresponding vertices. The two polygons are called the **bases** of the prism. If the connecting lines are perpendicular to the two planes, we say that the prism is a **right prism**. If the polygons are regular, we say that the prism is a **regular prism**. The examples in Figure 2 are all *right regular prisms.* The precise name of each prism is obtained from the name of the bases. For instance, if the bases are triangles, we call the prism a *triangular prism.* A nonright prism is shown below.

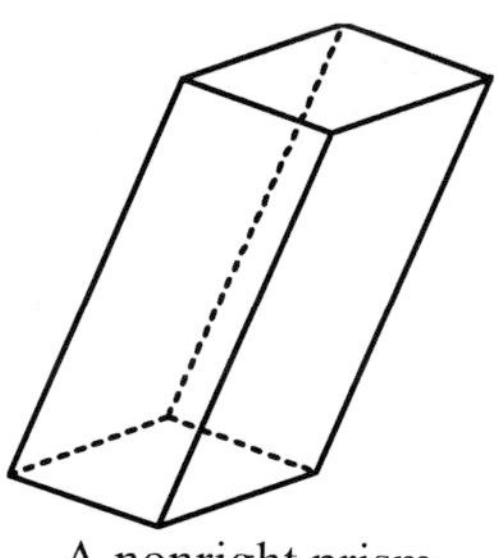

A nonright prism

***Figure 2 — Some right regular prisims***

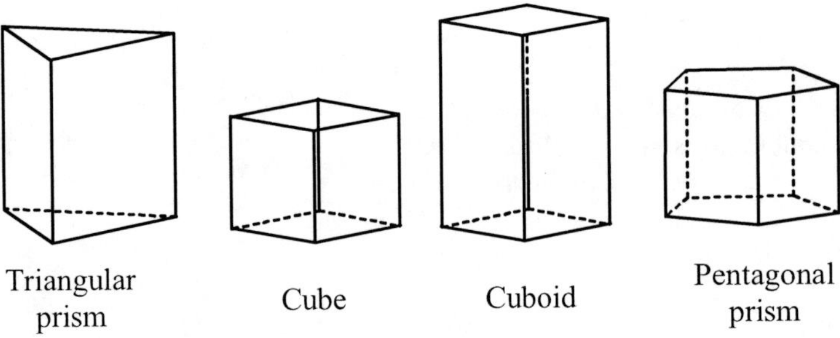


Figure 3 shows some examples of pyramids. A **pyramid** is a polyhedron formed by taking a polygon $P$ and a point $A$ that does not lie in the same plane as the polygon and then connecting each vertex of $P$ to the point $A$. The point $A$ is the **apex** of the pyramid and the polygon $P$ is the **base**. The pyramids in Figure 3 are **regular pyramids** since in each case the base is a regular polygon.

***Figure 3 — Some regular pyramids***

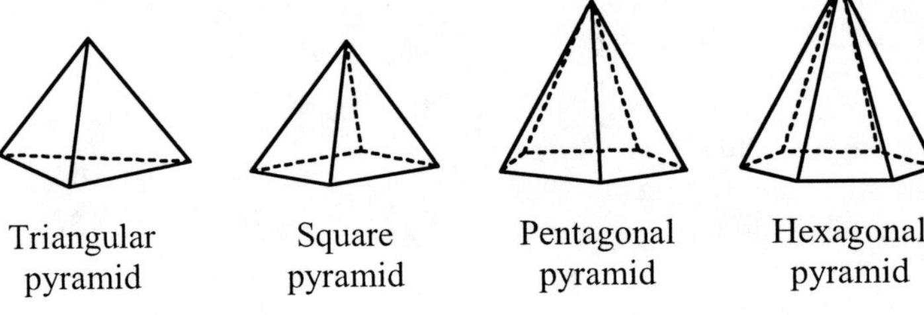


## Regular Polyhedra

Euclid, in Book XIII of the *Elements* (written about 300 B.C.), talked about the *regular polyhedra.* A **regular polyhedron** is a polyhedron in which all faces are congruent and which "looks the same" from all vertices and all edges. In other words, all *vertex figures* are the same and all *edge figures* are the same. Euclid showed, as we will in a moment, that there are only 5 regular polyhedra. These are shown in Figure 4. Plato (who lived around 400 B.C.) mentioned the regular polyhedra in his work *Timaeus* and for this reason, the 5 regular polyhedra are called the **Platonic solids**. You can see from the figure that a tetrahedron has 4 faces, a cube has 6 faces, an octahedron has 8 faces, a dodecahedron has 12 faces and an icosahedron has 20 faces.

***Figure 4 — The five regular polyhedra (Platonic solids)***

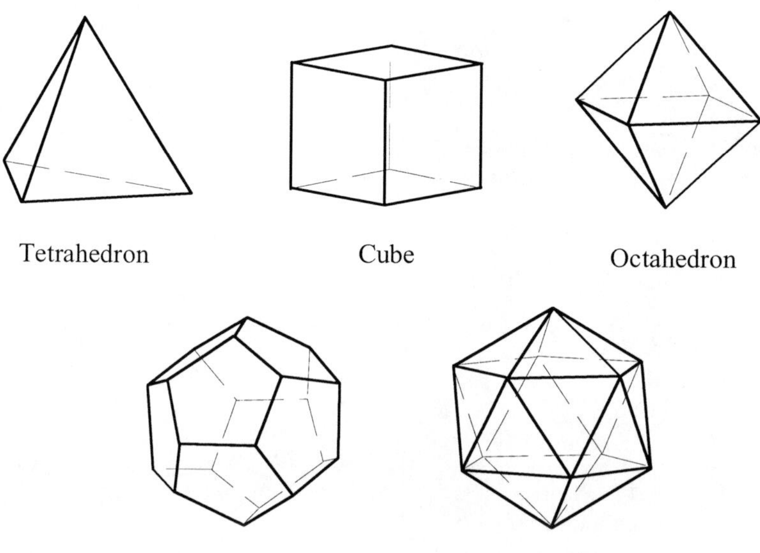


Figure 4 – The five regular polyhedra (Platonic solids)

Let us now consider the question of how we know that these five polyhedra are the only regular polyhedra. Suppose $P$ is a regular polyhedron and let $V$ be one of its vertices. Suppose also that $n$ regular $p$-gons meet at $V$ (and hence at any vertex). We recall from the previous section that the vertex angle of each face is $180(1 - 2/p)$ and since the sum of the vertex angles at $V$ must be *less than* 360° (why?), we have

$$n \cdot 180\left(1 - \frac{2}{p}\right) < 360$$

or

$$n\left(1 - \frac{2}{p}\right) < 2$$

Multiplying by $p$ and rearranging gives

$$\begin{aligned} n(p-2) &< 2p \\ np - 2n - 2p &< 0 \\ (n-2)(p-2) - 4 &< 0 \\ (n-2)(p-2) &< 4 \end{aligned}$$

Since there must be at least 3 faces at each vertex and each face must have at least 3 edges, we certainly have $n, p \geq 3$. Thus, $n - 2$ and $p - 2$ are both at least 1. Furthermore, in order for the product $(n-2)(p-2)$ to be less than 4, each factor must be less than 4 and so $n$ and $p$ must be less than 6.

Summarizing, we have

$$(n-2)(p-2) < 4, \quad 3 \leq n,p \leq 5 \qquad (1)$$

Now it is a simple matter to check all of the possibilities, to obtain the 5 solutions that correspond to the 5 Platonic solids. We leave that for you to do as an exercise.

We can get a very interesting and useful view of a convex polyhedron by moving our eye very close to one of the faces and peering into the figure, as shown below.

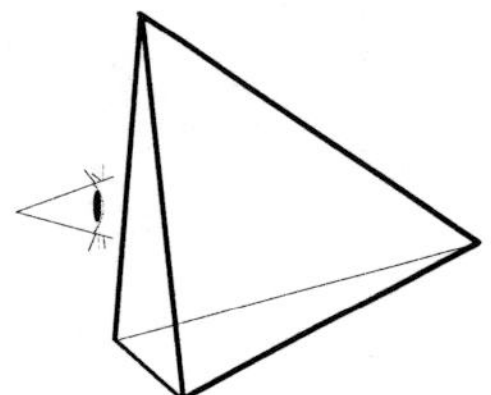

When doing this, we can see every vertex, edge and face as though they all lay in a single plane. This 2-dimensional picture of a polyhedron is called the **Schlegel diagram** of the polyhedron. Figure 5 shows the Schlegel diagrams of each of the five regular polyhedra. To help visualize how these are constructed, we have given the face that we peer through thicker edges and the face at the "back" has been shaded. Keep in mind that the polyhedra is 3-dimensional, but the Schlegel diagram is only 2-dimensional, that is, it is drawn in the plane. Note also that the face of the polyhedron that we peer through corresponds to the "outside" region of the Schlegel diagram.

***Figure 5 — The Schlegel diagrams for the Platonic solids***

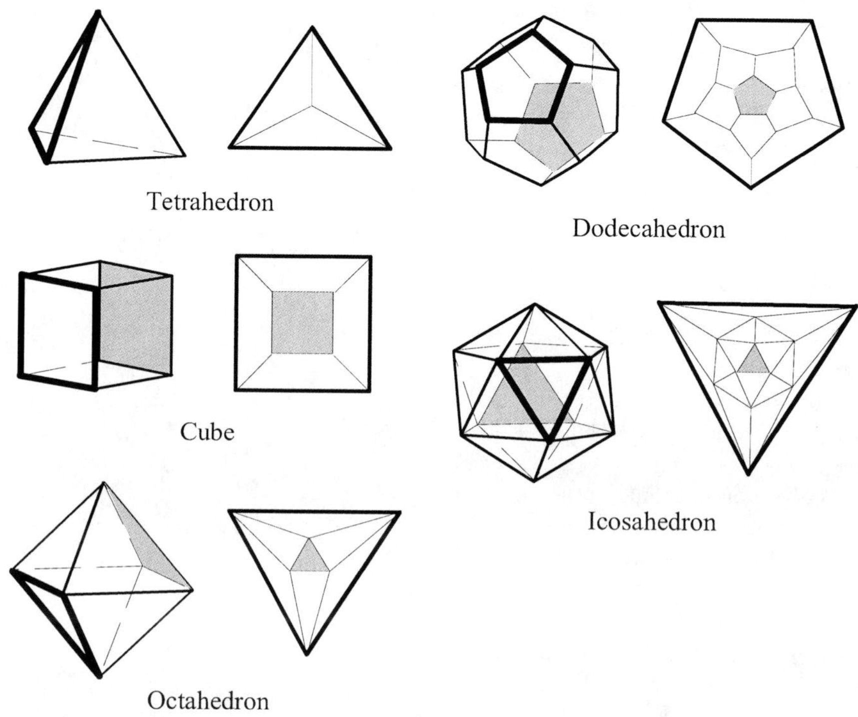


## Euler's Formula

There is a very interesting formula relating the number of faces, edges and vertices of a convex polyhedron. In order to derive this formula, let us look at a few examples. Consider the 3 polyhedra in Figure 6.

***Figure 6***

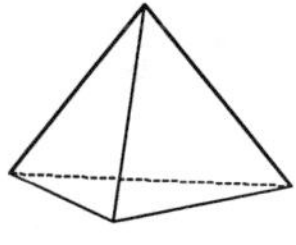 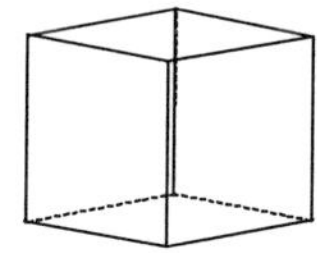 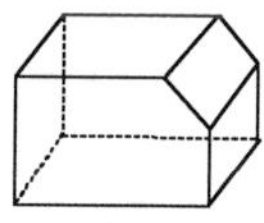

The first polyhedron has $f = 4$ faces, $e = 6$ edges and $v = 4$ vertices and we have

$$f - e + v = 4 - 6 + 4 = 2$$

The middle polyhedron in Figure 6 has $f = 6$ faces, $e = 12$ edges and $v = 8$ vertices and we also have

$$f - e + v = 6 - 12 + 8 = 2$$

Finally, the third polyhedron in Figure 6 has $f = 7$ faces, $e = 15$ edges and $v = 10$ vertices and we also have

$$f - e + v = 7 - 15 + 10 = 2$$

Thus, in each case, the number of faces minus the number of edges plus the number of vertices is equal to 2. This is no coincidence. It is true for *all* convex polyhedra and is known as **Euler's formula**.

**Theorem 1**

(**Euler's Formula**) If $P$ is a convex polyhedron with $f$ faces, $e$ edges and $v$ vertices then

$$f - e + v = 2$$

**Proof**. We can give a convincing argument to establish Euler's formula by looking at Schlegel diagrams. (This isn't quite a *proof*, but it is enough to get the idea of why Euler's formula holds.)

A Schlegel diagram is nothing more than a collection of points in the plane, representing the vertices of the polyhedron, connected by line segments, representing the edges of the polyhedron. The only restriction is that no two line segments intersect anywhere except at a vertex. (See Figure 5 for some examples.) In general, a collection of points that are connected by line segments in this manner is called a **planar graph**. In any case, instead of counting the number of faces, edges and vertices in the polyhedron itself, we can count these numbers in the Schlegel diagram, provided we remember that the exterior of the diagram counts as one face (the face through which we are peering).

The simplest planar graph is a triangle, which has $f = 2$ faces (interior and exterior), $e = 3$ edges and $v = 3$ vertices and so Euler's formula

$$f - e + v = 2 - 3 + 3 = 2$$

holds for this case. In fact, it is easy to see that Euler's formula holds for any planar graph that is composed of a single polygon with any number of sides. Such a graph has only 2 faces – one is the interior and the other is the exterior.

Now suppose that we add a new face to a planar graph with only 2 faces, as shown in Figure 7, where the new edges are drawn as broken lines.

***Figure 7***

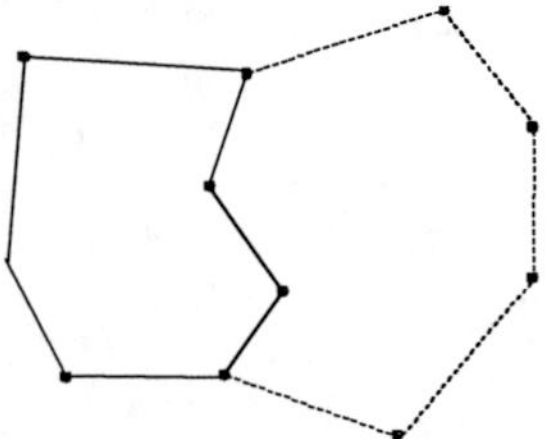

Before the new face was added (that is, without the broken edges) Euler's formula held. That is, $f - e + v = 2$. But the addition of the new face in Figure 7 has increased the number of faces by 1, the number of edges by 5 and the number of vertices by 4, for a *net gain* of $1 - 5 + 4 = 0$. Hence, Euler's formula holds even with the new face.

More generally, whenever we add a new face to an existing planar graph, as in Figure 8, the net gain in the formula $f - e + v$ is 0 and so Euler's formula continues to hold. Thus, Euler's formula holds for all planar graphs, hence for all Schlegel diagrams, and finally also for all convex polyhedra.

***Figure 8***

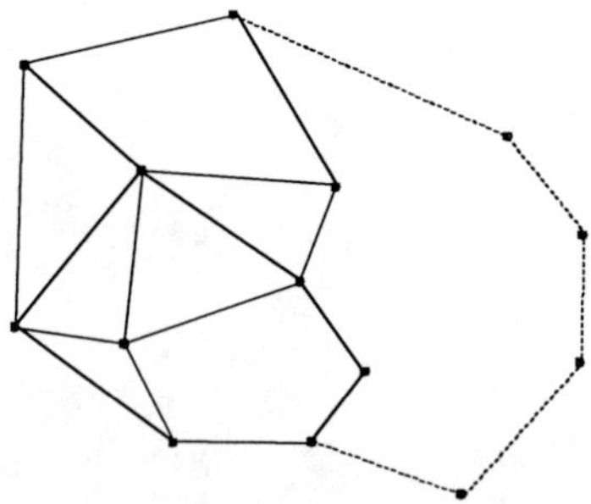

★

## EXERCISES

1. Draw a picture of a regular pentagonal prism that is not right regular.
2. A **parallelepiped** is a prism whose bases are congruent parallelograms. Draw a picture of a parallelepiped. Name the most famous special case of a parallelepiped.
3. The volume of a prism is the area of the base times the height of the prism. Find a formula for the area of a right regular triangular prism in terms of the length of one side of the base and the height of the prism.
4. Draw a right regular hexagonal prism, a nonright regular hexagonal prism, a regular nonright hexagonal prism and a nonright nonregular hexagonal prism.
5. Draw a picture of a regular septagonal pyramid.
6. Would it make sense to define a *right* pyramid? How would you do so?
7. Draw a picture of a cylinder. Is a cylinder a polyhedron? Justify your answer.
8. Draw a picture of a cone. Is a cone a polyhedron? Justify your answer.
9. Draw a polyhedron that is neither a prism nor a pyramid.
10. What do you suppose a *double pyramid* might be? Draw some examples.
11. Find the number of faces, edges and vertices in each of the Platonic solids.
12. Why must the sum of the vertex angles at a vertex of a convex polyhedron be *less than* 360°?
13. Find all of the solutions $n$ and $p$ that satisfy the conditions in the inequalities (1). Match the Platonic solids to the 5 solutions you have found.
14. Draw the Schlegel diagram of the polyhedron shown below, which is obtained from a cube by slicing off a portion of each corner. Peer through one of the octagonal faces for the diagram.

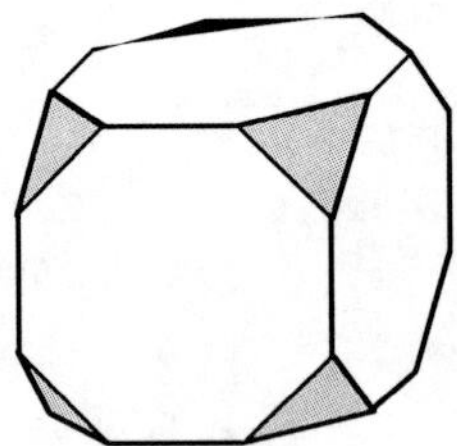

15. Show that Euler's formula holds for any planar graph that consists of a single polygon.
16. Verify Euler's formula for the following planar graphs.

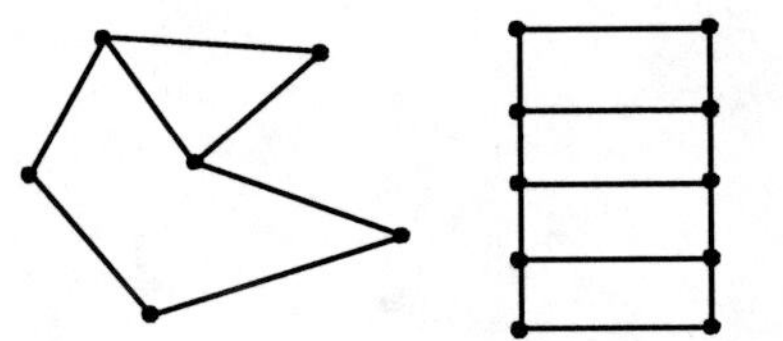

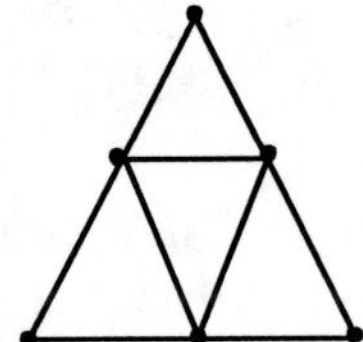

17. Verify Euler's formula for a polyhedron made from a cube by slicing off a small piece from one corner.
18. Verify Euler's formula for the following polyhedra. The second polyhedron is obtained from a cube by slicing off a portion of each corner.

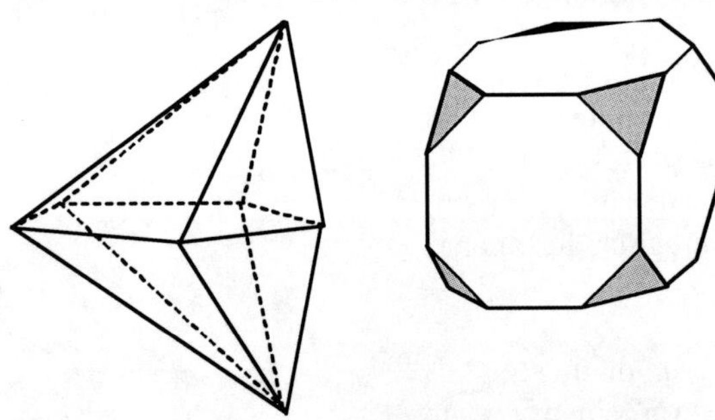

19. Imagine slicing off a small piece from the corners of an octahedron, as shown for a cube in the previous exercise. Verify Euler's formula for such a polyhedron.
20. The **dual** of a polyhedron $P$ is the polyhedron formed by connecting the centers of adjoining faces of $P$. Describe the dual of each of the 5 regular polyhedra.

# Chapter 2
# Fractals

## 2.1 Fractals

In this section, we give a brief introduction to the concept of a *fractal*. We will not give a precise definition of a fractal, since this would involve some very sophisticated mathematics. However, we can say that, roughly speaking, a fractal is a geometric figure that exhibits the same "irregularities" at all levels of magnification. This statement will become clearer as we give some examples of fractals.

One way to produce a fractal is to begin with a simple shape and transform it by a certain rule. Then the rule is applied to each piece of the resulting shape. We continue to apply the rule again and again. If we could do this forever, the result would be a fractal. Each time we apply the rule, we refer to the result as an **iteration**.

### The Koch Curve

Let us illustrate this process with one of the most famous fractals, called the **Koch curve**. We begin with a straight line segment.

Our rule is to divide this line segment into 3 equal length pieces and replace the middle segment with two new line segments of the same length, as shown below. (All 4 line segments have the same length.)

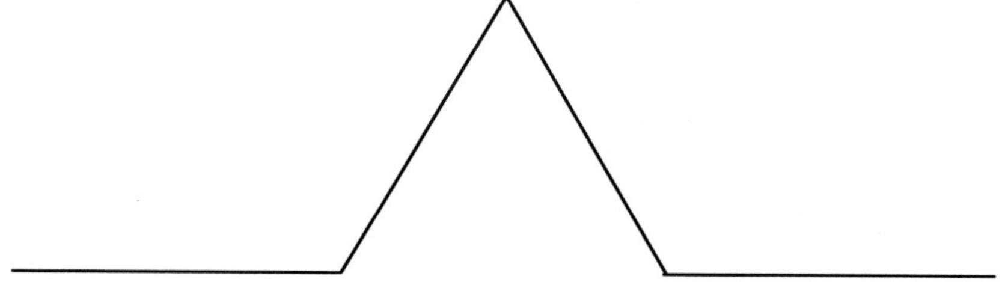

Repeating the rule requires that we perform the same replacement operation on each of the 4 line segments in the figure above, that is, we replace the middle third of each of the 4 line segments above by two new line segments. This gives

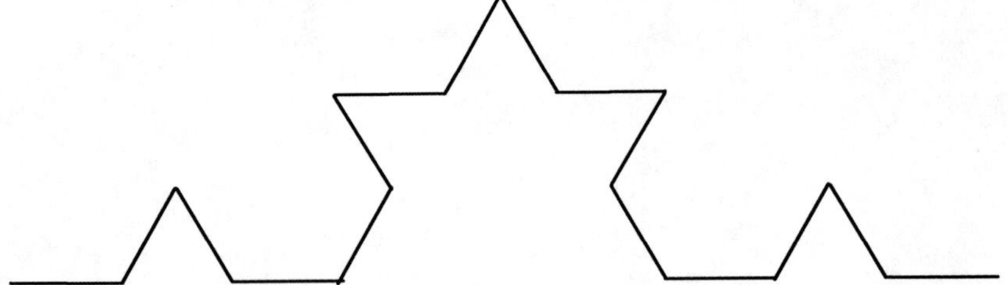

The process is continued indefinitely to produce the Koch curve. Of course, we cannot do that in practice, but here are a few more iterations. The beautiful emerging pattern is typical of the construction of a fractal.

## The Sierpinski Gasket

The Sierpinski Gasket is constructed beginning with a triangle

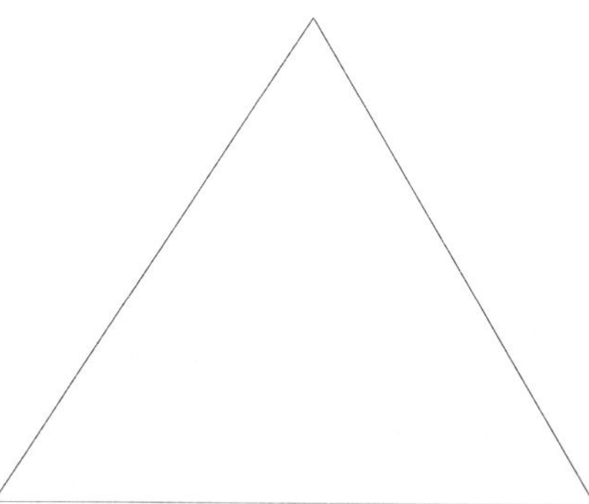

The rule for the Sierpinski gasket is to connect the three midpoints of the sides of this triangle, as shown below

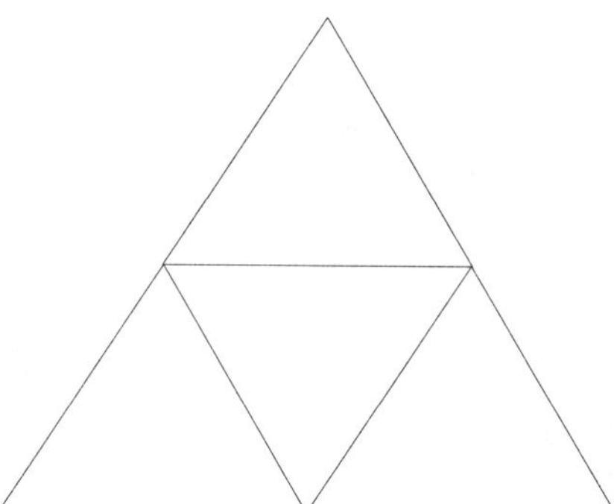

In each iteration, we connect the midpoints of the sides of each triangle. Here are some additional iterations

## The Heighway Dragon

To create the Heighway dragon, we begin with a straight line segment of length $d$

The rule for Heighway's dragon is to replace line segments by right angles. Thus, the line segment above is replaced by a right angle whose endpoints are a distance $d$ apart

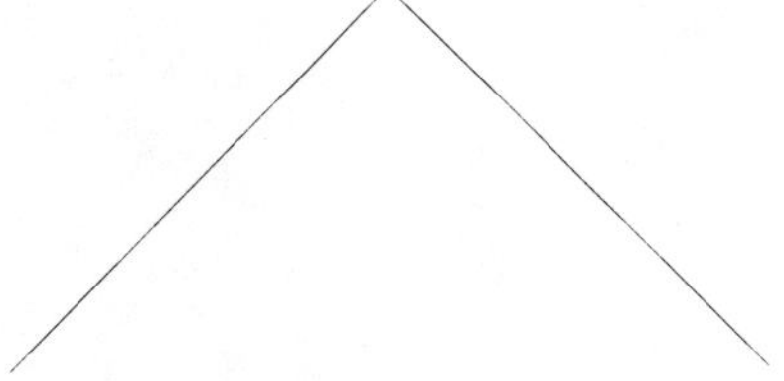

The next iteration requires that we replace each of the 2 line segments above by a right angle *pointing in opposite directions* as follows

We continue to replace line segments by right angles that alternate direction. Here are some additional iterations

## A Fractal Tree

For the fractal tree, we begin with a vertical line segment

The first iteration replaces this line segment by 3 new line segments, each one-half the length of the original, placed in the orientation shown in the figure below, where we also show the original line for comparison.

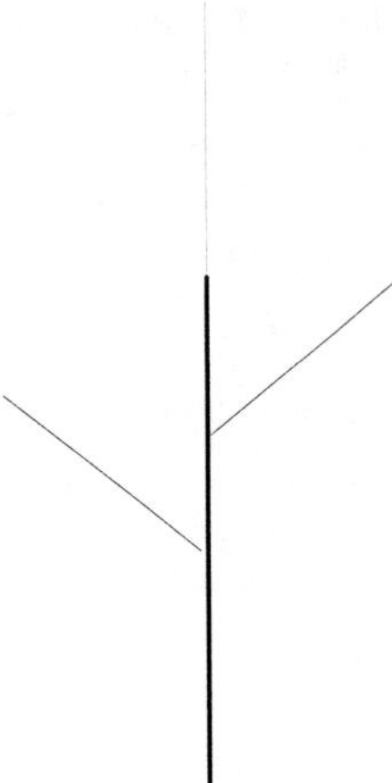

Thus, the first iteration is shown below below.

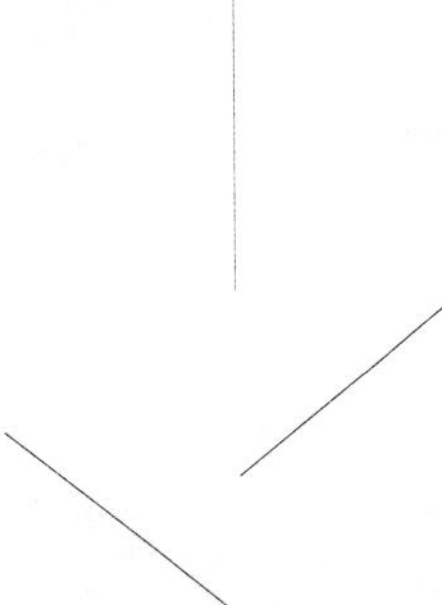

Each iteration replaces each line segment by 3 line segments as above. Here are the results of several iterations. The last figure shows what happens when you place all of the previous iterations together. It certainly looks like a real tree, doesn't it?

## The Dimension of a Fractal

As you no-doubt know, a line has dimension 1 and a plane has dimension 2. One of the characteristics of a fractal is that its "dimension" may lie *between* 1 and 2.

Actually, there are many different ways to define the dimension of a fractal. The one we will discuss is the simplest, and is called the *similarity dimension.* (The similarity dimension should not be confused with the *fractal dimension.*)

Consider the first iteration of a fractal construction. Typically, an object of a certain size is replaced by several congruent objects of a smaller size. If the length of the original object is $s$, the length of the smaller objects is $t$ and the number of smaller objects is $n$, we define the **similarity dimension** of the fractal to be

$$d = \frac{\log n}{\log\left(\frac{s}{t}\right)} \tag{1}$$

Let us compute the similarity dimension of some of the fractals we have constructed.

Consider the construction of the Koch curve and assume that we begin the construction with a line segment of length $s$. In the first iteration, a line segment of length $s$ is replaced by 4 line segments, each of which has length $t = s/3$. Hence, the similarity dimension of a Koch curve is

$$d = \frac{\log 4}{\log\left(\frac{s}{s/3}\right)} = \frac{\log 4}{\log 3}$$

This may be approximated using a calculator to get $d \approx 1.2619$. Thus, a Koch curve has greater dimension than a line, but less dimension than a plane. (Remember, this applies to the fractal itself. Each *iteration* has dimension 1.)

In the first iteration of a Sierpinski gasket, adding the new line segments has the effect of replacing the original triangle, which we can assume has height $s$, by 3 congruent triangles, each of which has height $t = s/2$. Hence, the similarity dimension of a Sierpinski gasket is

$$d = \frac{\log 3}{\log\left(\frac{s}{s/2}\right)} = \frac{\log 3}{\log 2} \approx 1.5850$$

For Heighway's dragon, the first iteration replaces a line segment of length $s$, say, by $n = 2$ line segments of length $t = s/\sqrt{2}$. This can be seen by applying the Pythagorean Theorem to the triangle pictured below

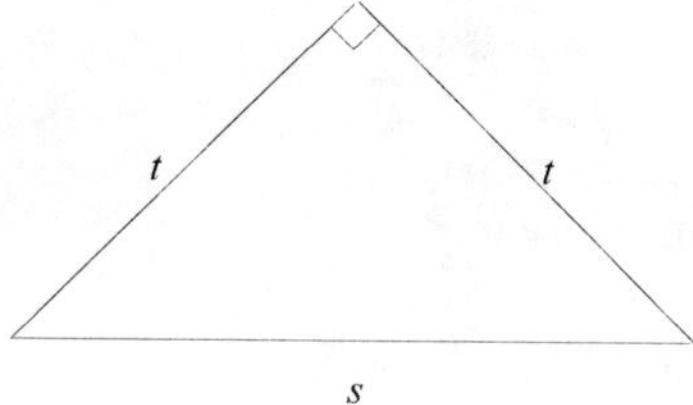


The base of this triangle is the original line segment, and the two sides of the triangle are the results of the first iteration. According to the Pythagorean Theorem,

$$t^2 + t^2 = s^2$$

or

$$2t^2 = s^2$$

$$t^2 = \frac{s^2}{2}$$

$$t = \sqrt{\frac{s^2}{2}} = \frac{s}{\sqrt{2}}$$

Thus, the similarity dimension of the Heighway dragon is

$$d = \frac{\log 2}{\log\left(\frac{s}{s/\sqrt{2}}\right)} = \frac{\log 2}{\log \sqrt{2}}$$

Now, according to the properties of logarithms (which you can find in any algebra book), we have

$$\log \sqrt{2} = \log 2^{1/2} = \frac{1}{2} \log 2$$

and so

$$d = \frac{\log 2}{\log \sqrt{2}} = \frac{\log 2}{\frac{1}{2} \log 2} = 2$$

Thus, the similarity dimension of Heighway's dragon is 2. We leave it as an exercise to compute the dimension of the fractal tree.

## Conclusions

Fractals are used in many different ways in applications to real world problems. As it turns out, the "regular irregularities" incorporated into fractals provide an excellent model for natural phenomenon. As Benoit Mandelbrot, who coined the term fractal, says in his book *The Fractal Geometry of Nature* (W.H. Freeman and Company, 1982) "Clouds are not spheres, mountains are not cones, coastlines are not circles, and bark is not smooth, nor does lightning travel in a straight line." When we look, for example, at a coastline from the air, we see a series of irregular inlets and outlets, measured in terms of miles. When we observe a smaller piece of the same coastline from the ground, we see a similar pattern of inlets and outlets, made up of large rocks and boulders, measured in terms of yards. When we observe an even smaller piece of the same coastline, we again see a similar pattern of inlets and outlets, perhaps made up of small stones, this time on a scale of inches. Even a microscopic portion of the coastline shows similar patterns, made up of grains of sand. Thus, the natural pattern of *self–similarity* persists at all levels of magnification, and this behavior is precisely what fractals are designed to model.

But fractals find their way into modern technology as well. For instance, fractals are being used to condense large amounts of data into a small space. This is referred to as *data compression.* Fractals can be used to compress data as much as 10000 to 1 and this promises to be a great boon to the communication industry.

**EXERCISES**

1. Find the similarity dimension of the fractal tree.
2. Construct a fractal by starting with a straight line segment. The rule is to divide the segment into five equal pieces and replace the second and fourth pieces each with two line segments of the same length as the one being replaced. The first iteration is shown below.

Draw the next iteration of this construction and compute the similarity dimension of the fractal.
3. Construct a fractal by starting with a straight line segment. The rule is to divide the segment into four equal pieces and replace the two center pieces each with two line segments of the same length as the one being replaced. The first iteration is shown below.

Draw the next iteration of this construction and compute the similarity dimension of the fractal.

4. Construct a fractal by starting with a straight line segment. The rule is to divide the segment into five equal pieces and replace the second piece and fourth pieces as shown below. (All resulting segments have the same length.) Note that the first "bend" is to the right and the second "bend" is to the left as you are traveling along the curve from left to right.

Draw the next iteration of this construction and compute the similarity dimension of the fractal.

5. Construct a fractal by starting with a straight line segment. The rule is to divide the segment into three equal pieces and replace the middle piece by four sides of a regular pentagon as shown below. (All resulting segments have the same length.)

Draw the next iteration of this construction and compute the similarity dimension of the fractal.

6. Construct a fractal by starting with a square. The rule is to divide each side of the square into three equal pieces and replace the middle piece by three sides of a smaller square as shown below. (All resulting segments have the same length.)

Draw the next iteration of this construction and compute the similarity dimension of the fractal.

7. Construct a fractal by starting with a square. The rule is to replace the square by eight smaller squares of equal size, as shown below.

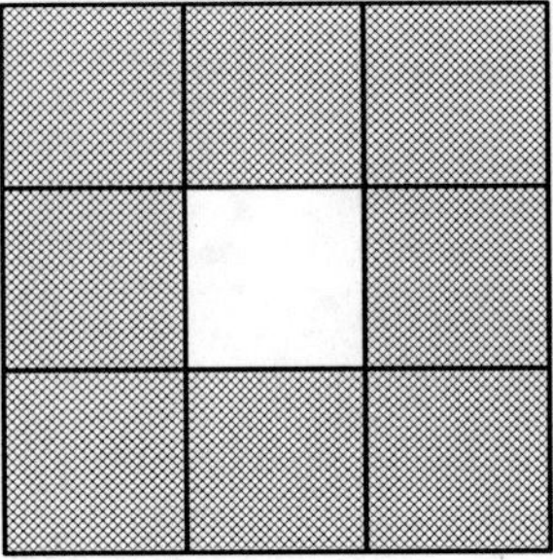

Draw the next iteration of this construction and compute the similarity dimension of the fractal.

8. Construct a fractal in a manner similar to that of Heighway's dragon, but instead of replacing the line segment by an angle of 90°, replace it by an angle of 120°. Draw four iterations of this construction and compute the similarity dimension of the fractal, which is known as the **120-degree dragon**.

9. Construct a fractal by starting with a single line segment. The rule is to replace each line segment by three equal length line segments as shown below. Note that the interior angles with the horizontal are 60° as shown in the figure.

Draw the next two iterations of this construction and compute the similarity dimension of the fractal, which is known as the **Sierpinski dragon**. Start the next iteration with the 60° angle facing to the *right*.

# Answers To Odd Numbered Exercises

**Section 1.1**

1. $f_5 = 641 \cdot 6700417$
3. These triangles are congruent since corresponding sides have equal length. Hence, the central angles $c$ are equal and since there are $n$ central angles, we have $nc = 360$, or $c = \frac{360}{n}$. Since the sum of the angles of a triangle is 180, we have $2d + c = 180$. Solving this for $2d$, which is the vertex angle, we get $2d = 180 - c = 180 - \frac{360}{n} = 180(1 - \frac{2}{n})$.
5. A square has 4 lines of symmetry, one through each pair of opposite vertices and one through the midpoints of each pair of opposite sides.
7. A regular $n$-gon has $n$ lines of symmetry. If $n$ is even, the lines go through each pair of opposite vertices and the midpoints of each pair of opposite sides. If $n$ is odd, the lines go through each vertex and the midpoint of each opposite side.
9. Yes, since rotating 360° does not change the figure. $\theta + n \cdot 360°$ for any integer $n$. (Clockwise rotation through a negative angle is counterclockwise rotation.)
11. $n \cdot 90°$ where $n$ is any integer.
13. The fundamental angle of rotation is $\alpha = \frac{360}{n}$, the central angle.
15. (a) Rotating through an angle $\theta$ brings $P$ to its original appearance. But this can be accomplished by performing two rotations, through an angle $\phi$ and then through an angle $\theta - \phi$. Since the first of these rotations brings $P$ to its original appearance, the second rotation (through the angle $\theta - \phi$) takes $P$ from its original appearance to its original appearance. Hence, $P$ has rotational symmetry through the angle $\theta - \phi$ as well. (b) Certainly, if $\alpha$ is an angle of rotational symmetry, so is any integral multiple of $\alpha$. On the other hand, suppose that $\theta$ is an angle of rotational symmetry. Then $\theta$ cannot be smaller than $\alpha$. (No angle of rotational symmetry, except 0, is less than $\alpha$.) Divide $\theta$ by $\alpha$, to get $\theta = a\alpha + \beta$ where $\beta$ is equal to 0 or is an angle less than $\alpha$. By part a), $\beta = \theta - a\alpha$ is an angle of rotational symmetry of $P$ and since it is less than $\alpha$, it must actually be equal to 0. Hence, $\theta = a\alpha$ is a multiple of $\alpha$.
17. Draw AB and construct a perpendicular bisector as shown on the left below. Then draw a circle through A with center O and mark point E. Finally, draw two circles through O with centers A and E, as shown on the far right below. Points A, O, E and F are the vertices of a square.

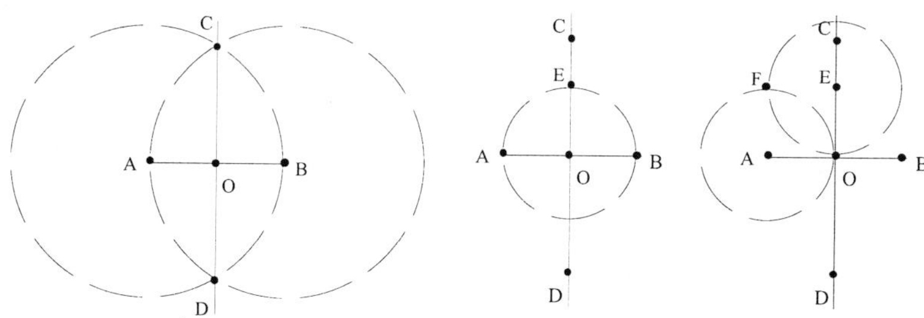

19.

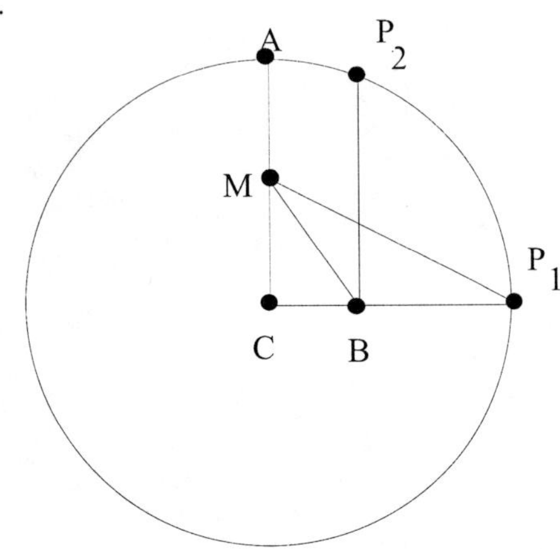


21. Construct an $rs$-gon. Connecting every $r$th vertex gives an $s$-gon and connecting every $s$th vertex gives an $r$-gon.
23. This can be done by drawing the two stars.
25. $d = 5, 7$
27. There are $n - 3$ star polygons since any number less than $n$ is relatively prime to $n$.

**Section 1.2**

5. Given the quadrilateral $Q$ in the center of this figure, we can surround $Q$ by quadrilaterals congruent to $Q$ by alternately rotating every other one 180°. The resulting 9 quadrilaterals can again be surrounded in the same manner. This can be continued indefinitely to produce a tiling of the plane by $Q$.
7. Since the vertex angle of a regular 43-gon measures $171\frac{27}{43}°$, the sum of the other angles in any possible vertex figure must be $188\frac{16}{43}$. But we have made a complete list of all possible sums of vertex angles that sum to $188\frac{4}{7}$ (which is greater than $188\frac{16}{43}$) in the text, and no such sum equals $188\frac{16}{43}$. Hence, the 43-gon cannot be used in a semiregular tiling.

9.

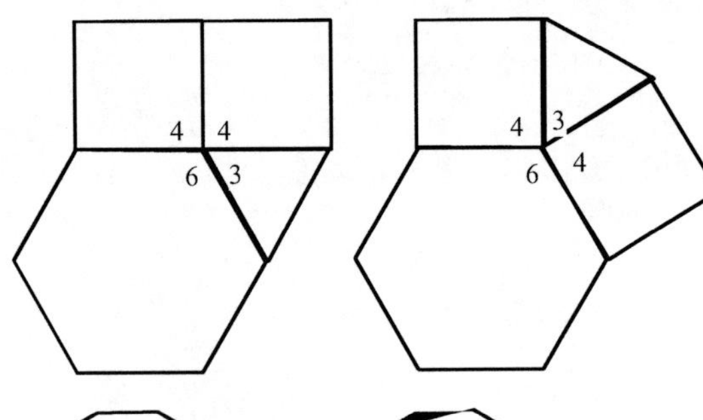


11.

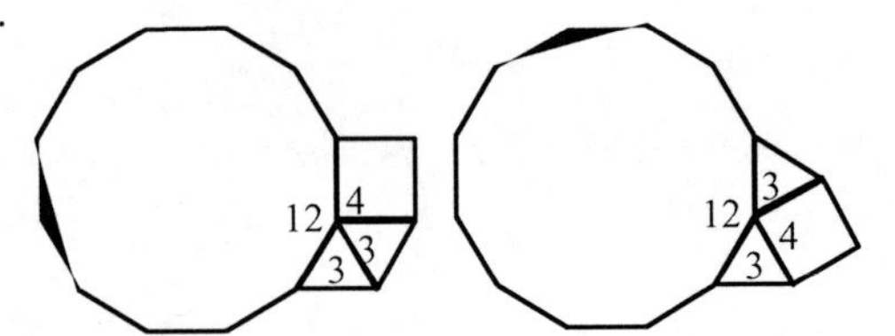

13. Referring to the figure below, the original vertex figure is of the type (12,4,3,3). In order for the vertex figure at $a$ to be the same type, the dodecagon A and the triangle B must be placed as shown. But then the vertex figure at $b$ will have a different type. Thus, we cannot obtain a semiregular tiling of the desired form.

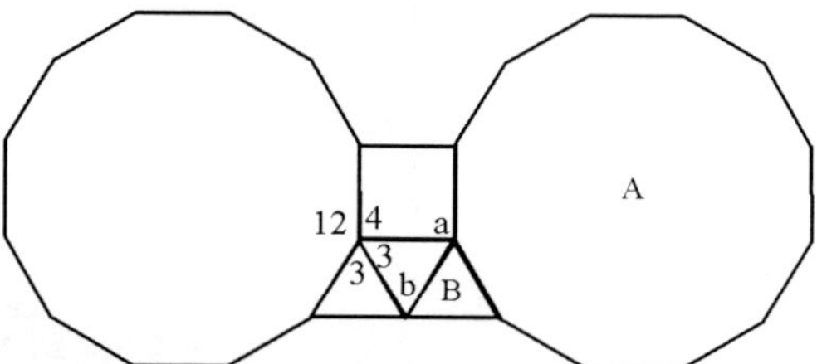


15. When the three sides of the triangle are surrounded by tiles of two different types, at least two of the surrounding tiles must have the same type. The vertex figure at which these two tiles and the triangle meet will then have two identical angles, which is not possible for a semiregular tiling based on any of these solutions.
17. Let $P$ be the pentagon in a tiling based on solution 20,5,4. When the five sides of $P$ are surrounded by tiles that are either dodecagons or squares, either two dodecagons or two squares will be adjacent. If these two polygons meet $P$ at vertex $v$, this vertex will have a different vertex figure than 20,5,4.
19. Referring to the figure below, in order for the vertex figure at $a$ to be the same as the original (6,6,3,3), we must place hexagon A and triangle B as shown. But then the vertex figure at $b$ will have a different type.

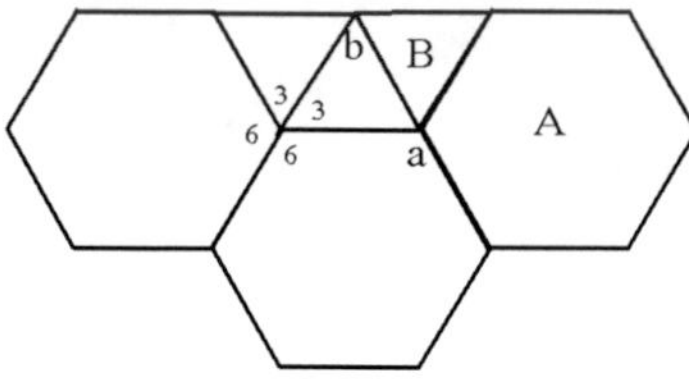


**Section 1.3**

1.

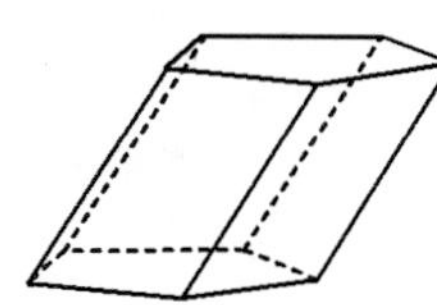

3. $V = \frac{\sqrt{3}}{2}sh$, where $s$ is the length of one side of the base and $h$ is the height of the prism.

5\.

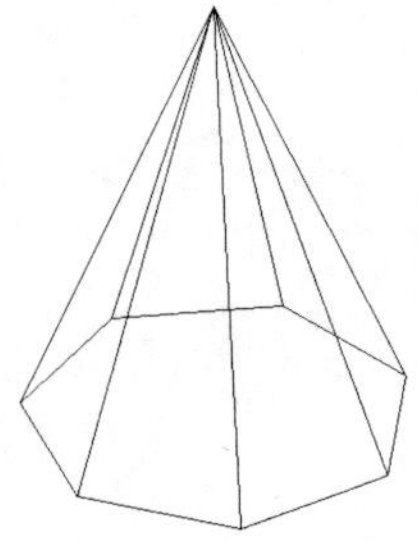

7\. No, since the base is not a polygon.

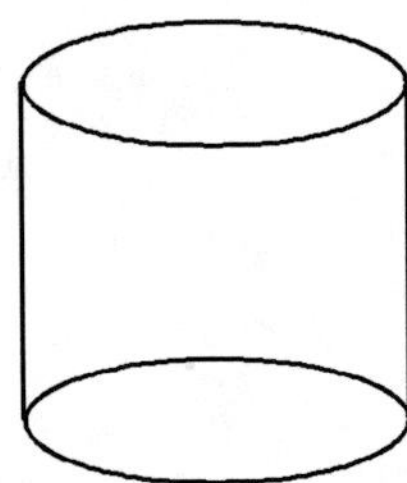

9\.

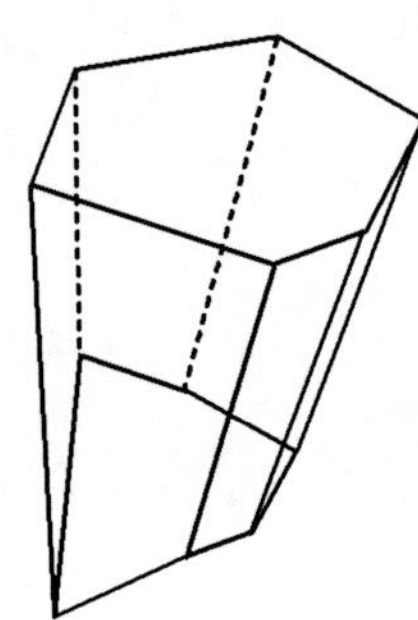

11. Tetrahedron: $f = 4$, $e = 6$, $v = 4$; Cube: $f = 6$, $e = 12$, $v = 8$; Octahedron: $f = 8$, $e = 12$, $v = 6$; Dodecahedron: $f = 12$, $e = 30$, $v = 20$; Icosahedron: $f = 20$, $e = 30$, $v = 12$
13. $n = 3$, $p = 3$, tetrahedron; $n = 3$, $p = 4$, cube; $n = 3$, $p = 5$, dodecahedron; $n = 4$, $p = 3$, octahedron; $n = 5$, $p = 3$, icosahedron.
15. If the polygon has $n$ sides, then it also has $n$ vertices. Since there are 2 faces, we have $f = 2$, $e = n$, $v = n$ and so $f - e + v = 2 - n + n = 2$.
17. For such a polyhedron, we have $f = 7$, $e = 15$, $v = 10$ and $7 - 15 + 10 = 2$.
19. For this polyhedron $f = 10$, $e = 20$, $v = 12$ and $10 - 20 + 12 = 2$.

**Section 2.1**

1. $d = \frac{\log 3}{\log 2} \approx 1.5850$
3. $d = \frac{\log 6}{\log 4} \approx 1.2925$
5. $d = \frac{\log 6}{\log 3} \approx 1.6309$

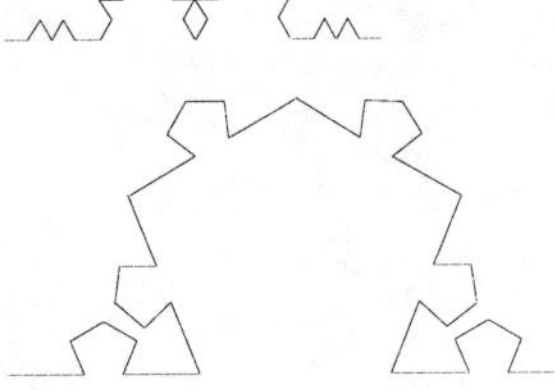

7. $d = \frac{\log 8}{\log 3} \approx 1.8928$

9.

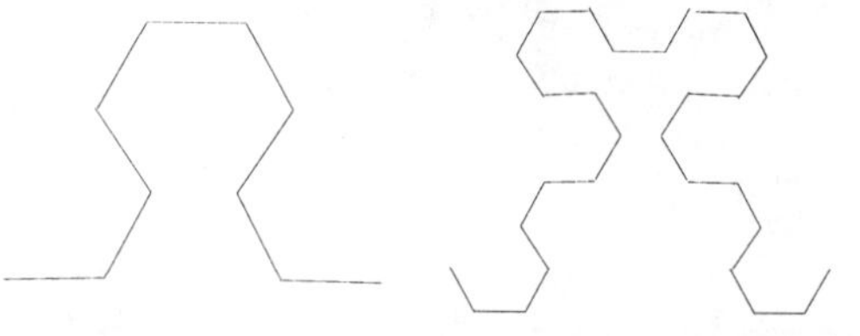